AF601917

Agro-based Adsorbents for Chromium Removal

The Author

Dr. Sohail Ayub is currently an Associate Professor in Environmental Engineering Section, Department of Civil Engineering, Aligarh Muslim University, Aligarh, U.P., India. He lectures extensively in the area of environmental engineering. With about two decades of various experience, in academic, industries, and consulting, his expertise in environmental engineering covers a variety of first hand practical experience both in field and in the laboratory.

Agro-based Adsorbents for Chromium Removal

Sohail Ayub

Department of Civil Engineering,
Faculty of Engineering and Technology,
Aligarh Muslim University,
Aligarh

2017

Scholars World

A Division of

Astral International Pvt. Ltd.

New Delhi – 110 002

Publisher's Note:

Every possible effort has been made to ensure that the information contained in this book is accurate at the time of going to press, and the publisher and author cannot accept responsibility for any errors or omissions, however caused. No responsibility for loss or damage occasioned to any person acting, or refraining from action, as a result of the material in this publication can be accepted by the editor, the publisher or the author. The Publisher is not associated with any product or vendor mentioned in the book. The contents of this work are intended to further general scientific research, understanding and discussion only. Readers should consult with a specialist where appropriate.

Every effort has been made to trace the owners of copyright material used in this book, if any. The author and the publisher will be grateful for any omission brought to their notice for acknowledgement in the future editions of the book.

Cataloging in Publication Data--DK
Courtesy: D.K. Agencies (P) Ltd. <docinfo@dkagencies.com>

Ayub, Sohail, author.
Agro-based adsorbents for chromium removal / author, Sohail Ayub.
pages cm
Includes bibliographical references.
ISBN 978-93-86071-29-3 (International Edition)

1. Agricultural wastes--India. 2. Sewage--India--Purification--Chromium removal. 3. Absorption. I. Title.

TD930.A98 2017 DDC 628.7460954 23

Published by : **Scholars World**
A Division of
Astral International Pvt. Ltd.
– ISO 9001:2015 Certified Company –
4736/23, Ansari Road, Darya Ganj
New Delhi-110 002
Ph. 011-4354 9197, 2327 8134
E-mail: info@astralint.com
Website: www.astralint.com

Acknowledgements

I first would like to thank my students for their comments, suggestions, and encouragement. Thanks to Professor Syed Iqbal Ali and Professor Nasim Ahmad Khan for their invaluable guidance and suggestions. Their combined influence has inspired me to complete this work. This book would not be accomplished without the support of Chairman Department of Civil Engineering AMU Aligarh. I am indebted to my colleagues and friends who were instrumental in the completion of this book.

Sohail Ayub

Preface

The technological, environmental and biological importance of adsorption can never be doubt. Its practical application in industry and environmental protection are of paramount importance. A need exists in the environmental engineering area for which present subject of electroplating wastewater treatment by agro-based adsorbents in a manner that can be readily followed and understand by engineers. The book has been essentially divided in to seven chapters and covers the entire treatment process.

Chapter one: This chapter covers the basic concept of chromium. Chemistry of chromium, application and disposal.

Chapter two: This consists of electroplating process. Sources of wastewater generation. Methods of plating wastewater treatment and disposal of metal plating waste.

Chapter three: Comprises three important information on adsorption process, interpretation of isotherms, types of adsorption systems and recent investigations.

Chapter four: Objective and its relationship with the previous.

Chapter five: This covers the material and methods and site description, sample collection, preparation and experimental procedure.

Chapter six: The chapter Results and Discussions consist of batch study results. Continuous system results as well as desorption. Hydrolysis test, SEM and comparison.

Chapter seven: This chapter contains conclusion and recommendations. Reference cited and appendix.

Sohail Ayub

Contents

List of Figures

List of Tables

Chapter 1

Introduction

1.1 General

The continually increasing demand for water for beneficial purposes has forced man to assess and examine water reuse technologies more seriously than ever. Regardless of origin, industrial wastewater, after proper treatment, represents another ample and reusable water source. Water pollution due to toxic heavy metals has been a major cause of concern to scientists and engineers. Several mishaps due to heavy metal contamination in aquatic environment increased the awareness about heavy metal toxicity. Public awareness for pollution caused by heavy metals is now worldwide. Among the many heavy metals, lead (Pb), mercury (Hg), cadmium (Cd), arsenic (As), chromium (Cr), zinc (Zn), and copper (Cu) are of most concern, although the last three metals are essential nutrients for animals and human. They find wide spread usage in industries. These metals enter the environment wherever they are produced, used or discarded. They are very toxic because as ions or in compound forms, they are soluble in water and may be readily absorbed into living organisms. After absorption, these metals can bind to vital cellular components such as structural proteins, enzymes, and nucleic acids, and interfere with their normal functioning. In humans, some of these metals even in small amounts, can cause severe physiological and health disorders.

The compound of chromium, especially hexavalent chromium, are known to be detrimental to human beings, plants, animals and aquatic life (Stomberg, 1984; Tondon, 1982). Over exposure of chrome workers to chromium dust and mists will cause irritation and corrosion of skin, respiratory track and probably lung carcinoma (Hayes, 1982). Ingestion may cause epigestric pain, nausea, vomiting, severe diarrhea and hemorrhage (Browing, 1969). On the other hand, trivalent chromium has been found to be an essential trace element in the human diet and deficiency in trivalent chromium has been linked with poor sugar metabolism (Katz, 1991).

Naturally occurring chromium is chiefly in the form of chromites (Cr_2O_3) or chrome-ore ($FeO-Cr_2O_3$). Chromium compounds are widely used in the electroplating of metal for corrosion resistance, plastic coating of surfaces for water and oil resistance, leather tanning and metal finishing, and in pigments and wood preservative etc. and consequently discharging hexavalent chromium bearing wastewater. Anthropogenic production of hexavalent chromium in smaller amounts is through drilling mud, rust and corrosion inhibitors, textiles and toner for copying machines (Sujatha, *et al.*, 1995). Hence it is a challenging task for the industrialists and environmentalists to dispose off waters containing heavy metals safely and effectively.

There are many techniques for the treatment of chromium bearing effluents, some of which are well established that have been in practice for decades such as precipitation, co-precipitation (Patterson and Passino, 1987), and concentration. These processes simply remove chromium from wastewater by reduction (Shen and Wang, 1995), coagulation and filtration. Although these technologies are quite satisfactory in terms of purging chromium and other heavy metals from wastewater, they produce solid residues containing toxic compounds whose final disposal is generally by land filling. It involves high costs and has possibility of ground water contamination. From environmental point of view, removing pollutants from liquid wastewater does not solve problem but transfer it from one phase (usually liquid) to another phase (usually solid). There is possibility that the presence of organic ligands and/or acidic conditions in the environment increases Cr (III) mobility, and also MnO_2, in the soil, could oxidize Cr (III) to more toxic and mobile Cr (VI) forms (Heary and Ray, 1987). Accordingly, the practice of land filling and land application of chromium contaminating sludge should be discouraged.

In recent years, the use of adsorption techniques for the removal of heavy metals has received global attention (Raji, *et al.*, 1998; Ajmal, *et al.*, 1995, 2000; Huang, *et al.*, 1975). Chemical contaminants at low concentrations are difficult to remove from the wastewater. Chemical precipitation, reverse osmosis and other methods become inefficient when contaminants are present in trace concentrations. The process of adsorption is one of the few alternatives available for such situations (Huang and Morehart, 1991). Several researchers have been working on the heavy metals removal. However, most of them used commercially available activated carbon. The high cost of activated carbon and its loss during the regeneration restricts its application. Thus there is need to undertake studies to substitute the costlier commercial activated carbon with the unconventional, low cost and locally available agricultural waste adsorbents (Bailey, *et al.*, 1999; Brown, *et al.*, 2000). India is an agricultural country and generates considerable amount of agricultural wastes such as sugarcane bagasse, coconut jute, nut shell, rice straw, rice husk, waste tea leaves, ground nut husk, crop wastes, peanut hulls, compost wastes etc. Studies on these materials could be beneficial to the developing countries and could be easily incorporated in development of appropriate technologies. Recently a few researchers have explored the possibility of using agricultural waste adsorbent for the Cr (VI) removal (Gang, *et al.*, 1999; Deo, *et al.*, 1992; Huang, *et al.*, 1975; Periasamy, *et al.*, 1991; Ayub, *et al.*, 1998, 1999, 2001, 2002; Drake, *et al.*, 1996; Shukhla, and Sakhardane,

1991; Weber, 1996; Chand, *et al.*, 1994; Siddiqui, *et al.*, 1994; Camino, *et al.*, 2000, Yaishya and Prasad, 1991).

The present study is to evaluate the heavy metal removal potential of agro-waste materials such as coconut shell, neem bark (Azadirachta indica) and raw sugarcane bagasse in the treatment of wastewaters. The effects of pH, contact time, adsorbent dose, concentration of metal, particle sizes and temperature were studied. The samples of the adsorbents were characterized before and after adsorption. The column studied were made to get the data to design a continuos system and to compare it with the batch performance. Electron microscopic technique was used to characterized the surface of the adsorbent. Various adsorbents were used to determine the removal potentials of different agricultural wastes. Thermodynamic nature of the process was also studied.

1.2 Chemistry of Chromium

Chromium is highly active transition metal which exists in a number of oxidation states which exhibit a wide range of stability. Thermodynamically, the reduced form of chromium, Cr (III), is the most stable in both acid and basic solutions as evidenced by the reduction potential (E°) diagrams presented in Figure 1.1.

$$\frac{1}{2}Cr_2O_7^{-2} \xrightarrow{1.3V} Cr(H_2O)_6^{+3} \xrightarrow{-0.4V} Cr(H_2O)_6^{+2} \xrightarrow{-0.9V} Cr^0 \qquad 1.1$$

(Acid) $Cr(H_2O)_6^{+3} \xrightarrow{-0.74V} Cr^0$

$$CrO_4^{-2} \xrightarrow{-0.13V} Cr(OH)_3 \xrightarrow{-1.1\ V} Cr(OH)_2 \xrightarrow{-1.4} Cr^0 \qquad 1.2$$

(Basic) $Cr(OH)_3 \xrightarrow{-1.34\ V} Cr^0$

Figure 1.1: Reduction Potential Diagram (*Source*: Shupack, 1991).

Positive E° values denote that the reaction will proceed spontaneously (ΔG°) to the right (reduction) under standard conditions, whereas, negative E° values indicate that the oxidized species is favored. Cr (III) is the most stable because it would require significant energy to reduce Cr (III) to Cr° or oxidize Cr (III) to Cr (VI) (Nieboer and Jusys, 1988).

The stability of the various chromium species is dependent upon the various reduction, oxidation and pH conditions. The stable domains for various chromium species in aqueous system as affected by the oxidation potential (E_h) and pH (Landrigan, 1975). Under the pH (< 3), temperature (20-30°C) and reducing - oxidizing conditions commonly found in industrial wastewaters, the predominant species are bicarbonate, $HCrO^-_4$, dichromate $Cr_2O_7^{-2}$ and Cr^{+3}. It is interesting to note that the divalent chromium ions, Cr^{+2}, may be found in extremely reducing environment. The concentration distribution between $HCrO^-_4$ and $Cr_2O_7^{-2}$ is largely governed by the total Cr (VI) present. The fraction of $Cr_2O_7^{-2}$ only becomes significant at high concentration of total Cr (VI) (Stumm, 1970).

1.2.1 Use of Chromium

Chromium is used to increase resistance and durability of metals by chrome plating. Chromium based pigments are used for floor covering products, paper, cement, and asphalt roofing. The other usages of chromium are in fabrication of alloys, colouring of glass, ferrous as well as non-ferrous, production and processing of insoluble salts, chemical intermediates; use in textile industry in dyeing, silk treating, printing, leather industry tanning and in photographic fixing baths. Catalysts for halogenation, alkylation, and catalytic cracking of hydrocarbons and fuel additives and propellant additives.

Chromium and its compounds are used in making special alloys protective coatings on metal; magnetic tapes; and pigments for paints, cement, paper, rubber, floor covering and other materials. In medicine, chromium compounds are used in astringents and antiseptics.

1.2.2 Disposal of Chromium

It has been routine practice to use land treatment or burial (sanitary landfill) methods to dispose off the chromium contaminated waste generated from the treatments. However, this practice is subjected to significant revision taking into consideration the underground water contamination. Disposal of chromium wastes from the electroplating operation can be divided into; (1) discharge of liquid effluents directly into the water courses, (2) discharge into municipal sewers.

The sludge generated by the precipitation process is generally voluminous and difficult to dewater. The removal of sludges from the wastewater is normally done by the gravity settling in a clarifier. Sludge generated by precipitation typically contains 0.5-3.5 per cent solids. Freshly precipitated sludge have the characteristics of a very low specific gravity, which makes effective settling difficult without long clarifier detention time.

Ion exchange is used to remove either trivalent or hexavalent chromium from wastewater. This method is preferred when chromium is to be recovered for reuse. It performs better in the pH range of 4.5 to 5.0. Ion exchange is an attractive method for the removal of small amount of impurities from dilute wastewaters, or the concentration and recovery of expansive chemicals from segregated concentrated wastewaters. The limited capacity of ion exchange systems mean the relatively large installations are necessary to provide the exchange capacity needed between regeneration cycles. It is an expansive methods and needs additional treatment unit for the spent resin regeneration. Wastewater treatment consisting of chromium reduction followed by ion exchange would normally not be economical.

Clarification, with or without sludge conditioning, is not completely effective in solid – liquid separation. Some of smaller floc will escape even the most effectively designed clarifiers. In order to treat increasingly stringent effluent standards, adsorption has shown potential for treating chromium bearing low concentration wastes as a polishing sludge free treatment technology.

Chapter 2

Electroplating Operations: Process Details, Wastewater Generation, Treatment Methods and Disposal

2.1 Introduction

Electroplating is an art of depositing a metal layer on to another metal surface by application of the direct current. The main reasons of electroplating are to give decorative finish and impart greater corrosion resistance. Decorative value of the objects is increased by depositing a metal onto the other cheaper metal by the electroplating. The usual types of plating are copper, chrome, nickel, cadmium zinc, silver and gold. The electroplating process cause environmental concern: (1) Discharge of contaminated wastewater, which contains heavy metals; (2) Some of heavy metals are quite toxic; (3) Solid waste arising out of the treatment of liquid waste (Naik, 2002). The metals used in electroplating are given in Table 2.1.

Electroplating processes are complex. However, an idea of how these processes work can be gained considering the electroplating of silver metal onto a copper object. The copper object and a piece of silver metal make up the two electrodes. They are dipped into a solution containing dissolved silver nitrate, $AgNO_3$ and other additives and direct current is passed. Electroplating deposits a smooth layer of silver metal. Silver ion in solution is reduced at the negative charged cathode, leaving a thin layer of silver metal. Silver ion is replaced in the solution by oxidation of silver metal at the silver metal anode (Stanley, *et al.*, 1993).

Following are the main objectives of metal plating:

A Metals'

i) To increase the resistance to corrosion of the plated metals.
ii) To increase the resistance to chemical attack and wear resistance.
iii) To improve physical appearance and hardness.
iv) To increase the decorative and commercial values of the metal.
v) To improve surface properties.

B. Non metal

i) For increase in strength.
ii) For prevention and decoration of surface of non-metals like wood.
iii) For making the surface conductive, and utilization of lightweight, non-metallic likes wood and plastic (Jain, *et al.*, 1996).

Table 2.1: Metals Used in Electroplating Bath (Neuhof, 1989)

Metals	*Alkaline with Cyanogens*	*Acid without Cyanogen*
Silver	Only cyanidic	–
Cadmium	Mainly cyanidic	–
Chromium	–	Mainly acid
Copper	Mainly cyanidic	–
Nickel	–	Mainly acid
Zinc	Mainly cyanidic	–
Brass	Mainly cyanidic	–

2.2 Electroplating Processes

The various steps likely to be carried out include cleaning, stripping, plating and rinsing. The process diagram of the chromium electroplating is shown in Figure 2.1.

The objects are cleaned first before electroplating. Surfaces of the objects are covered with the grease, dust and other impurities. Grease usually comes from machining, stamping, polishing and preservation stages. If the grease is of organic nature, it is removed by saponification with warm alkali. Petroleum and mineral oil greases cannot be removed by this method and trichloroethylene, benzene, gasoline and carbon tetrachloride are employed. But the most commonly employed agent for cleaning is commercially available emulsifier of alkalis available in the market as metal cleaners. These are usually mixtures of sodium carbonate, caustic soda, trisodium phosphate, sodium silicate, sodium cyanide and borax.

Removal of rust and scales is called pickling. It is done by treating the objects with sulphuric acid or hydrochloric acid in case of articles made of iron. The acids partly dissolve the scales and mechanical action removes remaining rust. The electrolytic method of stripping is also used because of its rapid action. In this

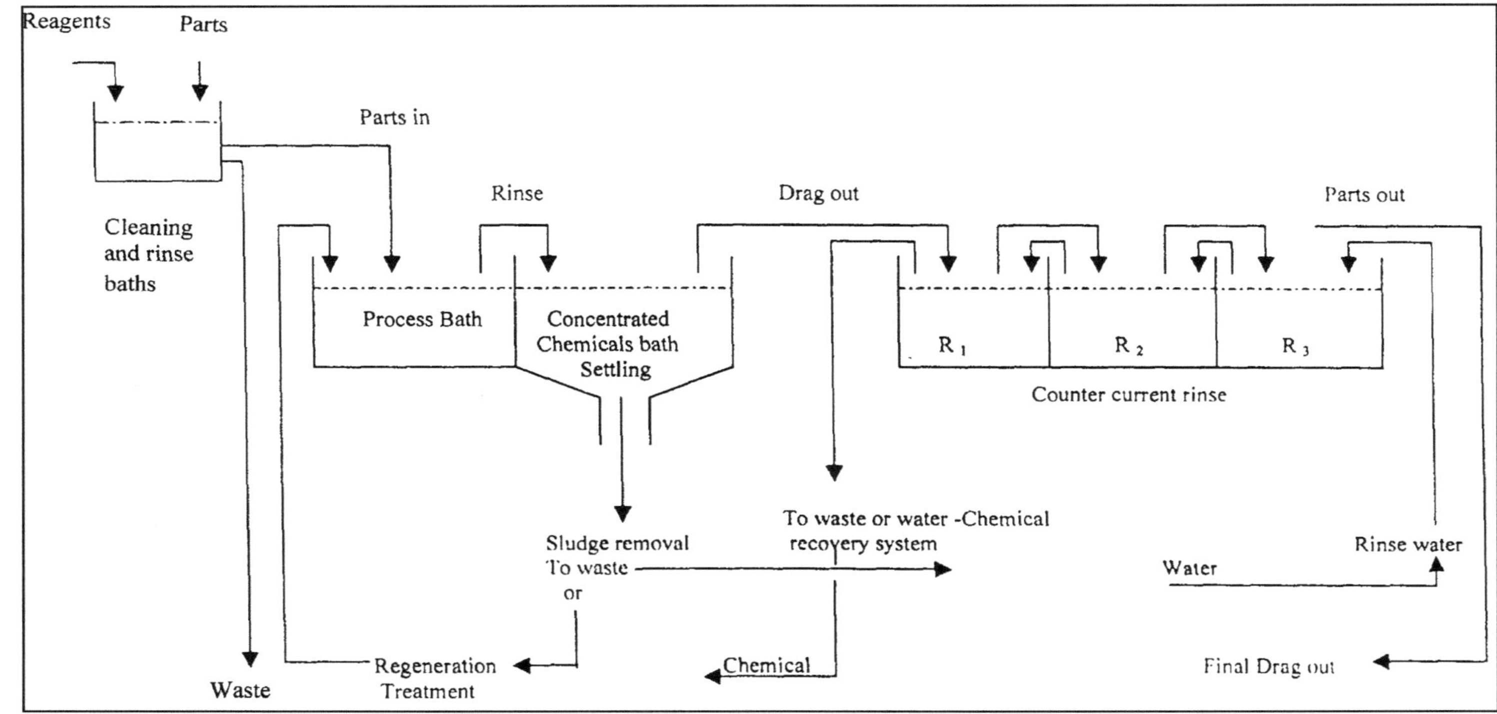

Figure 2.1: Generalized Schematic of the Integrated Treatment System (*Source*: Lancy, 1967).

method the material to be plated is made the anode, usually in an electrolyte baths. The end products in both the processes are essentially the same.

The pickled articles are placed in wooden or mild steel vats with special lining wherever necessary. The material being plated is made the cathode in an electrolytic cell. Plating baths are acidic in nature and generally contain sulphuric, hydrochloric or nitric acids. Where as alkaline baths containing sulphide, carbonate, cyanide and hydroxide are also used. The concentration of chemicals normally used in some of the common plating baths is given in Table 2.2.

Table 2.2: Typical Concentrations of Contaminants in Electroplating Baths and Rinse waters (IS: 7453-1974)

Sl.No.	*Bath Formulae*	*Bath Concentration*	*Rinse Concentration (Average 550l/h) (Drag out 3.4 l/h)*
1.	**Cadmium**	Cd - 19000 mg/l	Cd - 1.2 mg/l
	Cadmium cyanide 25 g/l	CN - 22000 mg/l	
	Sodium cyanide 30 g/l	pH - 9.2	PH - 8.3
2.	**Copper (Cyanide)**	CN - 20500 mg/l	CN - 8.6 mg/l
	Cuprous cyanide 22.5 g/l	Cu - 10400 mg/l	Cu - 1.34 mg/l
	Sodium cyanide 30 g/l	pH – 10.9	
	Sodium carbonate 10 g/l		PH - 8.2
3.	**Copper (Acid)**	Cu - 49000 mg/l	Cu - 2.3 mg/l
	Copper sulphate 220 g/l	pH - 0.2	PH - 6.3
	Sulphuric acid 60 g/l		
4.	**Chromium**	Cr - 140,000 mg/l	Cr - 30 mg/l
	Chromic acid 300 g/l	pH - 1.2	PH - 7.3
	Sulphuric acid 3 g/l		
5.	**Nickel**	Ni - 93000 mg/l	Ni - 6.3 mg/l
	Nickel sulphate 210 g/l	PH - 5.5	PH - 7.8
	Nickel chloride 60 g/l		
	Nickel citrate 30 g/l		
6.	**Zinc**	CN - 59000 mg/l	CN - 7.0 mg/l
	Zinc cyanide 50 g/l	Zn - 38000 mg/l	Zn - 6.2 mg/l
	Sodium cyanide 60 g/l	pH - 11	PH - 9.4
	Sodium hydroxide 65 g/l		

After plating has been done, the plated objects are rinsed with water. They are dipped in stationary water baths first, allowed to drain and then dipped in running water baths. Stationary baths are utilized to make solution for the plating operations while the running water baths are discharge into the drains. The quantity of drag out depends upon the nature of the solution, its temperature, shape of the material being plated and the time allowed for draining. Manual plants are known to have higher drag out losses than the automatic ones.

2.3 Sources of Wastewater

2.3.1 General

Waste from metal finishing operations particularly those from electroplating, are among the most toxic of industrial effluents. They contain poisonous constituents such as acids, many different heavy metals *viz.*, chromium (Cr), nickel (Ni), cadmium (Cd), zinc (Zn) copper (Cu) and cyanides. Electroplating waste is one of the major contributors to heavy metal pollution in surface waters. For this reason electroplating waste should not be discharged into surface streams unless treated and they should not be permitted entry into municipal sewers if biological treatments are employed. Pretreatment of metal finishing wastes at the source, to reduce the concentration of the toxicants below the environmentally acceptable levels should be provided before discharging into the industrial wastewater treatment facility or municipal sewage.

The main sources of wastewaters in the plating operations are bath solutions and rinse waters they are distinctly different in volume and chemical quality. Bath solutions from vats are highly concentrated and are seldom discharged. Rinse waters are comparatively much more dilute but form the bulk of the wastes.

2.3.2 Cleaning Solutions

Essentially objects are cleaned before carrying out electroplating. Several cleaning reagents are used for the purpose. The washout from the cleaning vats may also contain heavy metal contaminants in the wastewaters. There are cleaners of various types and are generally prepared according to the manufacture specifications. They are spilled out during the drag out operation. The rinse vats are generally continuos in flow. The pH remains in the neutral range. Effluents from the cleaning vats is normally combined with other streams to be treated together as the wastewater from the electroplating plant.

2.3.3 Spent Alkaline and Rinse Waters

They include all the spent alkaline solutions containing suspended solids, soaps, grease, and globules of oil. The frequency of discharge of these wastes varies from plant to plant and depends to a large extant on the grease protection given to the metallic surfaces before they are received in the shop. The pH of these wastes when discharged intermittently is usually very high (around 12.0). They are generally held in steel tank for controlled discharge or for blending with acid wastes for lowering the pH. The pH of alkaline wastes from dip and rinse operations varies from 8.8 to 9.8.

2.3.4 Acid Pickling and Rinse Waters

They contain mostly ferrous sulphate and strong spent acids usually with pH below 2.5. The frequency of discharge of pickling wastes is much more than the cyanide concentrates. Pickled and acid dipped articles are washed in vats provided with continuos flow of water. This rinse water is acidic and has pH in the range of 4.5 to 5.6. This operation is normally not continous and the rinse waters are stored in bitumen or acid proof brick- lined tank for controlled discharge required in blending operations.

2.3.5 Cyanide Concentrations

They arise from cyanide vats, dips, drip from the articles and rinse operation and contain fairly large quantities of cyanide that is very much harmful to life even in very low concentration. The discharge is usually controlled by holding the wastes in a tank. Detention time gives an opportunity for pretreatment before they discharged for further treatment. Rinse waters do not pick up large quantities of cyanide during the washing operation. These are continuous as compared to those from the vats and dips. The vats are emptied once in every 6 to 8 weeks and the contents are very rich in cyanide. The composition of cyanide bath liquors is given in Table 2.3.

Table 2.3: Composition of Cyanide Bath Liquors [Neuhof, 1989]

		Composition
1.	Copper electrolyte	
	Copper	14-20 g/l
	Cyanide	22-27 g/l
	Tartaric acid	0-15 g/l
2.	Zinc electrolyte	
	Zinc	20-35 g/l
	Cyanide	30-50 g/l
	Sodium hydroxide	40-90 g/l
3.	Brass electrolyte	
	Copper	10-30 g/l
	Zinc	5-30 g/l
	Cyanide	25-80 g/l
	Sodium hydroxide	0-15 g/l
4.	Cadmium electrolyte	
	Cadmium	10-20 g/l
	Cyanide	40-50 g/l
5.	Silver electrolyte	
	Silver	3-30 g/l
	Cyanide	20-40 g/l
	K_2CO_3	0-100 g/l
6.	Gold electrolyte	
	Gold	0.5-4 g/l
	Cyanide	2-10 g/l
	Na_2HPO_4	70 g/l

2.3.6 Chromate Wastes

The bulk of the chromate wastes originate from chromium plating, anodizing, electroplating solutions and dip solutions like passivating dips, bright dips, etc.

and small portions arise from rinsing operations of metals treated with chromate solutions. The plating vats are normally not discharged. The strength of the plating solutions is maintained by the addition of chromic acid and sulphuric acid. The main source of chromium in the wastes is from the drag out and washing operations. Chromium concentration varies from 3 to 30 mg/l depending upon the care with which the plating operations are carried out. The pH of the rinse waters is generally in the neutral range and rarely goes below 5.5. The waste is stored for a regulated discharge after proper treatment to reduce hexavalent chromium to trivalent state.

2.3.7 Metal Wastes from other Plating Wastes

They include rinse waters from copper, zinc, nickel, cadmium and lead vats. All these metals seldom occur together because the baths are never used all at a time and the operations are staggered to meet the demand. However, a combination of these baths is practiced resulting in the appearance of several of these metals at the time in the mixed rinses. The metals are present in soluble ionic form and most of them are extremely toxic.

2.3.8 Floor Washes

Occasionally the plating shop is washed by running water to clean the floor. During the plating operations, some spilling and splashing occurs from all the baths and the wash water thus possibly contain cyanide and almost all the metals that are used for plating. The concentration of these toxicants depends on the carelessness of the operations and varies widely.

2.3.9 Regeneration Wastes

Most of the plating operations need demineralised water to prepare vat solutions. After exhaustion, the strong acid cation exchange resin is regenerated with 2 to 5 percent sulphuric acid or 5 to 8 percent hydrochloric acid solution and the weak base anion exchange resins with 4 to 6 percent sodium hydroxide or 6 to 8 percent sodium carbonate solution. During the regeneration and rinse operation, the excess regenerates drain out as a waste. The regeneration frequency depends on the size of the columns and the quantity of demineralised water required.

2.3.10 Composite Wastewater Concentration

The volume and characteristics of the wastewater very considerably from one plating shop to the other. This is primarily because of varying efficiencies in handling operations. Typical characteristics of industrial wastes resulting in plating operations are given in Table 2.4. The physico chemical characteristics of Morris Bajaj electroplating waste are given in Table 2.5. The Morris Bajaj lock factory is situated in the industrial area of Aligarh, UP (India) where electroplating of locks and other building materials is done (Ajmal, *et al.*, 1992).

2.4 Methods of Plating Waste Treatment

The effluent from the electroplating contains highly toxic chemicals. Discharge of untreated waste into the municipal waste or industrial waste would be a deleterious effect on biological sewage treatment processes due to the presence

Table 2.4: Typical Characteristics of Industrial Waste Resulting in Plating Operations [IS: 7453-1974]

Sl.No.	Waste	Flow L/h	pH	Basicity mg/l	Cyanides mg/l	COD mg/l	Total Solids, mg/l	Suspended Solids, mg/l	Other Items, mg/l
1.	Cleaning solutions (rinse waters)	450-680	7.8-8.4	300-650	0	290-350	960-1120	610-720	Fe: 2.3-3.1
2.	Cyanide concentrations (rinse waters)	450-680	9.2-9.9	800-1700	0.3-21.2	25-42	430-600	23-35	Cu: 2.8-9.7 Ca: 0.3-0.8 Zn: 1.8-2.4
3.	Acid Pickling rinse waters	900-1360	4.5-5.5	28-43	–	–	450-590	76-141	Fe: 80-110
4.	Spent Alkali rinse waters	2700-3650	8.8-9.8	180-410	–	300-550	800-1350	–	–
5.	Chromate rinse waters	1360-2270	5.5-6.8	31-56	–	–	460-750	–	Cr (VI): 3-30 Pb: 0.1-0.7
6.	Copper (cyanide) rinse water	450-680	–	210-340	7.3-11.6	–	–	–	Cu: 1.1-1.4
7.	Copper (acid) rinse waters	450-680	6.1	97-110	–	–	–	–	Cu: 1.8-3.1
8.	Nickel rinse waters	450-680	7.4-8.3	187-299	–	–	–	–	Ni: 4.8-8.7
9.	Cadmium rinse waters	450-680	8.0-8.8	191-320	3.2-4.6	–	–	–	Cd: 1.8-3.1
10.	Zinc rinse waters	450-680	8.9-9.8	210-405	5.4-9.0	–	–	–	Zn: 1.8-3.1
11.	Floor wash waters	–	7.6-8.0	170-230	0.1-0.3	–	350-480	65-79	Cu: 0.05-0.1 Zn: 0.3-0.5 Cd: 0.1-0.2 Ni: 0.1

of acids, alkalis, and toxic metallic ions such as Cr (VI), Cu (II), Zn (II) etc. these impurities could inhibit the biological processes in the municipal wastewater treatment plant or industrial wastewater containing high level of BOD requiring bioxidation.

Table 2.5: Physico-chemical Characteristics of Morris Bajaj and Sigma Engineering works Electroplating Wastewater (Ajmal, *et al.*, 1992)

1.	Colour	Yellowish	Yellowish
2.	Temperature (ºC)	25	22
3.	PH	6.0-8.5	8.7
4.	Turbidity (FTU)	240	150
5.	Total solids	1125.0	1980
6.	Suspended solids	215.0	315
7.	Dissolved solids	910.0	1665
8.	Sulphate	1050.0	995.0
9.	Chloride	185.0	216.0
10.	Fluoride	1.7	1.90
11.	Nitrate	1.5	1.40
12.	Na	175.0	172.0
13.	K	24.9	26.3
14.	DO	1.5	1.90
15.	BOD	12.5	10.2
16.	COD	95.5	78.86
17.	Cr (VI)	52.0	46.8
18.	Ni	8.9	10.1
19.	Cu	3.0	2.8
20.	Zn	65.5	58.5
21.	Cd	0.05	0.04

All units except colour, temperature, pH and turbidity are expressed in mg/l.

The purpose of the treatment is to remove the cyanides, hexavalent chromium, nickel, cadmium, zinc, cupper, iron, grease and oil so that the treated effluents meet the standards to discharge them in a water body. Volume of water including variations in flows, load of contaminate including variations in concentration and regulatory agency's requirements are some of the important variables to be studied before suggesting a treatment method. Foremost principle of environmental engineering *i.e.* waste minimization is adopted in order to curb the quantity of waste produced. The effluent quantity generation could be minimized by good shop keeping, adoption of healthy operational practice and occasional auditing of waste generation from the various sources.

The treatment of plating wastes by chemical and physical means are designed primarily to accomplish; (1) removal of cyanides; (2) removal of chromium and other heavy metals; (3) removals of oils and greases. There are several methods of cyanides removal from plating wastes. The selection of cyanides removal process depends upon the flow rate, hindrance by other impurities, fluctuation in flow or cyanide concentration as each method is specifically suitable for a situation. Cyanides removal method could be for the continuous flow or for the batch process. Caustic soda and chlorine are to oxide the cyanides by the following reaction.

$$2NaCN + 5\,Cl_2 + 12\,NaOH = Na_2 + 2\,Na_2\,CO_3 + 10\,NaCl + 6\,H_2O \quad 2.1$$

2.4.1 Precipitation

Once the cyanides are neutralized, the chromium is removed by the process of reduction and precipitation. The hexavalent chromium generally remains as chromic acid or chromates. The chromium reacts with reducing agents such as $FeSO_4$, SO_2, or $NaHSO_3$. After the reduction is complete, and alkali (usually lime slurry) is added to neutralize the acid and precipitate the trivalent chromium

$$H_2Cr_2O_7 + 6FeSO_4 + 6H_2SO_4 = Cr_2\,(SO_4)_3 + FeSO_4 + 7H_2O \quad 2.2$$

$$Cr_2\,(SO_4)_3 + 3Ca\,(OH)_2 = 2Cr\,(OH)_3 + 3CaSO_4 \quad 2.3$$

$$Fe\,(SO_4)_3 + 3Ca\,(OH)_2 = 2Fe\,(OH)_3 + 3CaSO_4 \quad 2.4$$

Treatment of other metals, oil and grease bearing wastes by neutralization and precipitation usually involves recombining the waste with previously oxidised cyanide and reduced chromium wastes for subsequent and final treatment. If the combine waste is acidic, alkali usually 5 – 10 per cent lime slurry, is added to neutralize and precipitate the metals. The floc produced large and quite heavy and hence the velocity of flow is decreased after the adequate flocculation has occurred. The waste is then followed to settle. Sludge is removed and usually lagooned since it is the most economical treatment for them slow drying relatively innocuous, metal sludges.

Chrome, nickel and copper acid type plating solution may be reclaimed from the rinse tank while evaporation in glass lined equipment, and the concentration solution returned to the plating systems. The water condensed from the steam is then used in the rinse tank following the plating tank, to eliminate build up of natural water salts. This process has proved effective for recovering valuable metal salts. The high initial cost for equipment is more than recovered not only because of waste treatment but also because of recovery of metals especially when volumes of metallic waste are large.

A typical metal bearing waste contains other metal ions as well as chromium to be removed by precipitation and because optimum pH of precipitation varied for different metal hydroxide, an average waste stream pH of about 8 often seems to achieve the best overall results on a waste containing several metal ions. Soda ash may be used to precipitate trivalent chromium although the most commonly employed chemical for precipitation is lime. The chromium content of electroplating waste may be was reduced from 140 - 1.0 mg/l by neutralization and precipitation with lime at pH 7-8. A coagulant aids was employed to improve the chromic hydroxide precipitate removal.

Chromate reduction by sodium metabisulphite below 0.1mg/l is followed by waste neutralization with lime to pH 9.0. Precipitation of chromic and other metallic hydroxide and settling is a promising technique. Trivalent chromium level below 0.2 mg/l is achieved by such treatment. Settling properties of the precipitate could be improved by addition of an ionic polyelectrolyte.

2.4.2 Ion Exchange

Ion exchange is more a concentration than final treatment technique. During treatment the ion exchange resins exchange its ions for those in the wastewater. This process continues until the solution being treated exhausts the resin exchange capacity. The exhausted resin is generated by another solution which replaces the ions given up in the ion exchange operation, thus converting the resin back to its original composition and yielding a concentrated regenerated brine. The regenerated brine is then treated more easily than the original volume of wastewater. In some cases, wastewater to be treated by ion exchange is generally filtered to remove suspended solids which could mechanically clog the rinse bed. Oils, organic wetting agents, brighteners, which might foul the resins, must also be removed perhaps by passage of wastewater through adsorbent carbon filters.

Successful ion exchange treatment has been reported for chromium wastewater primarily as the hexavalent chromate and dichromate ions (Yuronis, 1968; Rothstein, 1958; Richardson, *et al.*, 1968). The procedure utilizes an ionic exchange resin, normally regenerated with concentrated sodium hydroxide. The elute is sodium chromate, from which purified chromic acid may be recovered by removal of sodium in a cationic exchange system. Copper containing wastewater has been successfully treated by cation exchange, (Pinner, *et al.*, 1971; Schore, 1972) as have lead, (Liebig, *et al.*, 1943) mercury, (Cheremisinoff, *et al.*, 1972; Gardiner and Munuz, 1971) nickel (Recents and Stromquist, 1952; Heldorn and Keller, 1958) and other metals wastewater. Ion exchange is an attractive method for the removal of small impurities from dilute wastewater, or the concentration and recovery of expensive chemical from segregated concentrated wastewaters. Ion exchange is capable of achieving very high levels of copper removal, particularly from low concentration waste. Copper removal by ion exchange from 1.02 mg/l to less than 0.03 mg/l has been reported. There are many ion exchange resins with high specificity for the cadmium. However ion exchange method is unsuitable for recovery of cadmium from mixed cadmium cyanide solution. In most of industrial wastewaters, chromium exist in the hexavalent form and its recovery is preferred direct ion exchange treatment of chromate or dichromate is carried out. The different resins are employed to recover each form of chromium. In the ion exchange system pH is a critical factor with proper pH adjustment. Successful ion exchange treatment of a metal finishing waste has been reported to meet a chromate effluent standard of 0.05 mg/l (this chromate level is equivalent to a chromium (VI) concentration of 0.03 mg/l).

2.4.3 Reverse Osmosis

In reverse osmosis higher pressure than the natural flow pressure is applied in order to make the water to flow in the reverse direction. Most of the development work and commercial utilization of the reverse osmosis process, especially for desalination and water treatment and recovery, has occurred during the past two decades. Reverse osmosis functions most efficiently on dilute solutions. The greater the ionic strength solution, the higher the pump pressure and horsepower required to produce reasonable permeate flows. Reverse osmosis is therefore well suited to the concentration of dilute nickel plating rinse waters. Osmosis is a natural phenomenon involving the inclination of water to pass through a semi permeable membrane from the weak solution side to the strong solution side.

2.4.4 Adsorption Process

A solid surface in contact with a solution tends to accumulate a surface layer of solute molecules because of unbalance of surface forces. Chemical adsorption results in the formation of a monomolecular layer of the adsorbate on the surface through forces of residual valence of the surface molecules. Physical adsorption results from molecular condensation in the capillaries of the solid. In general, substances of the highest molecular weight are most easily adsorbed. There is a rapid formation of an equilibrium interfacial concentration, followed by slow diffusion into the capillary pores in the particles. The overall rate of adsorption is controlled by the rate of diffusion of the solute molecules within the capillary pores of the carbon particles. The rate varies reciprocally with the square of the particle diameter, increases with increasing concentration of solute, increases with increasing temperature, and decrease with increasing molecular weight of the solute.

It is generally feasible to regenerate spend adsorbent for economic reasons. In this process, the object is to remove from the carbon pore structure the previously adsorbed materials. The modes of generation are thermal, stream, solvent extraction, acid or base treatment, and chemical oxidation.

Adsorption studies are significant as they can be utilized in detection, determination and treatment of inorganic and organic pollutants and pesticides (Anonymous, 1972; Agro and Culp, 1972; and Maruyama, *et al.*, 1972). The use of the adsorption process for the removal of the pollutants by the use of the solid adsorbents has been considerably developed recently. It has been found to be very useful and successful in many fields such the purification process, water treatment process and analytical methods.

In recent years the adsorption techniques for the removal and the recovery of the heavy metals has received a great attention. Activated carbon has been used as an effective adsorbent for removing the organic and the inorganic contaminants from water, wastewater and air samples. It has been very effectively and successfully used with different adsorbent for the removal of various pollutants. Activated carbon has been found to be an effective adsorbent for the removal of many organic as well as inorganic substances present in the water. However the use of activated carbon posses economic problems and limits its large-scale use for the abatement of heavy metal pollution in developing country like India. Different agricultural waste adsorbent has been proved quite useful due to its low cost and wide availability *viz.* sugarcane bagasse, coconut jute, nut shell, rice straw, rice husk, waste tea leaves, ground nut husk, crop wastes, peanut hulls, fertilizer wastes etc. (Saeed, *et al.*, 1999; Gang, *et al.*, 1999; Deo, *et al.*, 1992; Huang, *et al.*, 1975; Periasamy, *et al.*, 1991; Ayub, *et al.*, 1998, 1999, 2001; Drake, *et al.*, 1996; Shukhla and Sakhardane, 1991; Weber, 1996; Chand, *et al.*, 1994; Siddiqui, *et al.*, 1994; Camino, *et al.*, 2000)

2.5 Disposal of Metal Plating Wastes

The disposal of wastes from plating operations can be divided into; (1) Discharge of liquid effluents directly into watercourse; (2) Discharge into municipal sewers. Often electroplating effluent disposal is not carried out with the care and attention merited by environmental considerations. The deliberate and willful dumping of solutions creates most problems. In case metal finishing unit effluents are discharge in to sewers or water coarse it may have serious consequences. It may interfere with ; (1) biological treatment facility; (2) lethal to aquatic organisms and plant; (3) threat to the drinking water. The discharge limitations for metal contaminants are shown in Table 2.6.

Table 2.6: Discharge Limitations for Selected Metal Contaminants (Federal Register, 1981 and Indian Standards 1993)

Parameters	*Max. Conc. Allowed for Discharge to Sanitary Sewer, mg/l Before Treatment*	*Effluent Guidelines mg/l*	
		Federal Register	*Indian Standard*
Cadmium	0.7	0.1	2.0
Copper	2.7	0.5	3.0
Chromium (Hexavalent)	–	0.1	0.1
Total Chromium	4.0	0.5	2.0
Cyanide	1.0	0.1	0.2
Zinc	2.6	0.8	5.0
Nickel	2.6	0.5	3.0
Lead	0.4	–	0.1

The enforcing agencies, administrative policies and legislative interference have overcome the problem of indiscriminate discharge of metal plating effluents in all the developed countries. However, there are many cases of serious interference with biological treatment processes at sewage treatment plants receiving municipal effluents in admixture with such wastes. Careful planning and design is effective in reducing the pollution load. With rapid industrialization, increasing concern for the protection of environment and the introduction of stringent water pollution control effluent disposal legislations have affected the Indian electroplating industry. The effect may be more pronounced for operation of medium and small size electroplating units which are unlikely to have any infrastructure for the safe disposal of their wastewater. There are more than 50,000 electroplating units in India mostly scattered in the urban areas (Rai, *et al.*, 1998).

Issues like use of alternative less hazardous chemicals, pretreatment of effluents followed by final treatment, waste segregation and recovery, safe disposal of sludge etc., have not been promoted properly in India. All possible efforts should be made and implement measures should be enforced to reduce the wastewater volume and level of its contamination. Efforts should be made to prevent the pollution with the emphasis on source reduction through operational improvement.

The sludge generated by precipitation process is usually voluminous and difficult to dewater. The removal of sludges from wastewater is normally done by gravity settling in a clarifier. Schwoyer and Luthingir, (1972) have reported that sludge generated by precipitation collected in clarifiers typically contain 0.5 - 3.5 per cent solids. Freshly precipitated sludge have the characteristics of a very low specific gravity, which makes effective settling difficult without long clarifier detention time or additional treatment. Clarification, with or without sludge conditioning, is not completely effective in solid – liquid separation. Some of the smaller floc will escape even the most effectively designed clarifiers. Sand filters are commonly used as finer devices for metal plating. Stone, (1967) has reported that the filtration would reduce a post clarifier chromium concentration ranging from 1.3 – 4.6 mg/l down to a range of 0.3 – 1.3 mg/l. This final residual was primarily in the soluble chromium (VI) form. Even filtration will not result in complete removal, particularly when a fraction of metal remains in soluble form.

In order to meet increasingly stringent effluent standards. Adsorption system is employed as polishing process for the complete removal of metals.

Chapter 3

Adsorption: Adsorption and Review of Recent Investigations

3.1 General

The process of adsorption is a tool for the environmental remedial and has many applications for treatment of municipal and industrial wastewater. Activated carbon is the most widely used commercial adsorbent and in general, has a great capacity for the adsorption of organic molecules such as dyes, phenols, detergents, cresols, or other toxic or nonbiogradable materials including heavy metals.

The surface of solids or liquids possess properties which give rise to the process of uptake of atoms, molecules or ions from the substances in contact. This phenomenon is known as adsorption. Thus adsorption is a change in concentration of given species on the surface layer in comparison to that in the bulk phase. The system which retains a given specie at the interface is known as the adsorbent and the retained species are referred to as adsorbate. The adsorption process can occur on all interfaces. In general there are five types of interfaces: solid-gas, solid-liquid, solid-solid, liquid-gas, and liquid-liquid. The solid-liquid interface has applications in various biological, industrial, and natural processes.

The adsorption is classified on the basis of interactions involved between the atoms, molecules or ions of adsorbate and adsorbent. (1) Physical adsorption; (2) Chemical adsorption; (3) Exchange adsorption, are the three basic kinds of adsorption processes.

3.1.1 Physical Adsorption

In this type of adsorption the van der Waals forces are responsible for the adsorption of species on the surface of adsorbent. Physical adsorption is reversible and rapid; a condition of equilibrium is established between adsorbed and unadsorbed adsorbate species. Molecules adsorbed by physical adsorption are held to the adsorbent surface by weak van der Waals forces of attraction or perhaps by p-bonding under certain conditions (Snoeyink and Weber, 1967). Adsorbed molecules are free to move on the surface of the adsorbent and adsorption is assumed to be multilayered with each new layer of molecules forming on the top of previously adsorbed layers. The amount adsorbed is predominant at low temperature and this type of adsorption is characterized by a low heat of adsorption.

3.1.2 Chemical Adsorption

This type of adsorption takes place due to the action specific 'chemical forces' *i.e.*, those involved in the transfer or sharing of electrons between an adsorbent and adsorbate species. Because of specific interaction, chemical nature of the adsorbent has a significant role. The chemical reactions on the surface of the adsorbent may be either exothermic or endothermic in nature. The increase in temperature will decrease or increase the extent of adsorption. The enthalpy change is usually in the range of 20 – 100 Kcal mol^{-1}. It is observed that the bonds formed between the surface of adsorbent and adsorbate species are almost as strong as those existing in the stable stoichiometric compounds (Snoeyink and Weber, 1967). Desorption of molecules is difficult and possible at elevated temperatures.

3.1.3 Activated Adsorption

These are made from variety of materials including wood, lignin, bituminous coal, lignite, and petroleum residues. Granular activated carbons produced from medium volatile bituminous coal or lignite has been most widely applied to the treatment of wastewater. Activated carbon has specific properties depending on the material source and the mode of activation. Mostly granular carbons from the bituminous coal have a small pore size, least surface area, and the lowest bulk density. Adsorptive capacity is the effectiveness of the carbon in removing desired constituents from the wastewater (Eckenfelder, 1988).

3.2 Theory of Adsorption

A solid surface in contact with a solution tends to form a surface layer of solute molecules because of imbalance in surface forces. Chemical adsorption results in the formation of monomolecular layer of the adsorbate on the surface through forces of residual valence of the surface molecules. Physical adsorption results from molecular condensation in the capillaries of the solid. In general, substances of the higher molecular weight are easily adsorbed. There is a rapid formation of an equilibrium interfacial concentration, followed by slow diffusion into particles. The rate varies reciprocally with the square of the solute. Morris, *et al.* (1964) found the rate of adsorption to vary as the square root of the time of contact with the adsorbent. The rate also depends upon pH because of change in surface charges.

The adsorptive capacity of a adsorbent for a solute will be dependent on both the adsorbent and the solute. Most wastewaters are highly complex and vary widely in the adsorbability of the compounds present.

3.2.1 Adsorption Phenomena

Adsorption is taking up of molecules by the external or internal surface of solids or by the surface of liquids. Adsorption occurs on the surface because of attractive forces of the atoms and molecules that makeup the surfaces and the adsorbate. When species are adsorbed from a liquid onto a surface, the adsorption process occurs at the solid liquid interface, and reactions occurring at the interface determining the rate and extent of adsorption. The transfer of solute from the solution to adsorbent continues until the concentration of the solute remaining in solution is in equilibrium with the concentration of solute adsorbed by the adsorbent. When equilibrium is reached, the transfer of solute stops and the distribution of solute between the liquid and solid phases is measurable and is well defined.

The equilibrium distribution of solute between the liquid and solid phases is an important property of adsorption systems and helps to define the capacity of a particular system. The kinetics of the system which describes the rate at which this equilibrium is reached are of equal importance. The rate of adsorption determines the detention time required for the treatment. Three distinct steps take place for adsorption to occur.

The adsorbed molecules are transferred from the bulk phase of the solution to the surface of the adsorbent particle. In doing so, it passes through a film of solvent that surrounds the adsorbent particle. This process is referred to as film diffusion. The adsorbate molecules then are transferred to an adsorption site on the inside of the pore. This process is referred to as pore diffusion. The adsorbate must become attached to the surface of the adsorbent, *i.e.* be adsorbed. Many factors influence the rate at which adsorption reactions occur and the extent to which a particular material can be adsorbed. The rate of adsorption is controlled by either film diffusion or pore diffusion, depending on the amount of agitation in the system. If relatively little agitation occurs between the adsorbent particle and the fluid, the surface film of liquid around the particle will be thick and film diffusion will likely be the rate-limiting step. If adequate mixing is provided, film around the adsorbent become thinner and the rate of film diffusion will increase to the point that pore diffusion becomes the rate-limiting step.

Particle size and surface area are important properties of an adsorbent. The size of particles influence the rate at which adsorption occurs: adsorption rates increase as particle size decrease. The total adsorptive capacity of an adsorbent depends on its total surface area. The size of an adsorbent particle does not have a great effect on total surface area, since most of the surface area lies within the pores of an adsorbent particle.

Solubility of the adsorbate also plays an important role during adsorption. Soluble compounds have a strong affinity for their solvent and thus are more difficult to adsorb than the less soluble compounds. However, there are exceptions, since many compounds that are slightly soluble are difficult to adsorb, whereas some very soluble compounds may be adsorbed readily (Benefield, *et al.*, 1982).

Molecular size would logically be important in adsorption, since the molecules must enter the micro pores of carbon particles so as to be adsorbed. Research studies have shown that within a homologous series of aliphatic acids, aldehydes, or alcohols adsorption usually increases as the size of the molecule become greater (Hassler, *et al.*, 1974). This can partly be explained by the fact that the forces of attraction between a adsorbent and a molecule are greater the closer the size of the molecule is to the size of the pores in the adsorbent (Culp, *et al.*, 1971). Adsorption is strongest when the pores are just large enough to permit the molecules to enter. Most wastewaters contain a mixture of compounds representing many different sizes of molecules. In this situation there would appear to be a danger of molecular screening, *i.e.*, large molecules blocking the pores to prevent the entrance of small molecules. However, the irregular shape of both the molecules and the pores, as well as the constant motion of the molecules, prevents such blockage from occurring (Culp, *et al.*, 1971). Furthermore, the greater mobility of the small molecules allows them to diffuse faster and to enter the pores ahead of the large molecules.

Adsorption process is also affected by pH and temperature. The pH at which adsorption is carried out has a strong influence on the extent of adsorption. This is partly due to the fact that hydrogen ions themselves are strongly adsorbed and partly that pH influences the ionization, and thus the adsorption, of many compounds. Organic acids are more adsorbable at low pH, whereas the adsorption of organic bases is favoured by high pH. The optimum pH for any adsorption process is determined by laboratory testing. The temperature at which an adsorption process is conducted will affects both the rate of adsorption and the extent to which adsorption occurs. Adsorption rates increasing with increased temperature and decrease with decreased temperature. However, since adsorption is an exothermic process, the degree of adsorption will increase at lower temperature and decrease at high temperatures.

3.2.2 Adsorption Rate Kinetics

Results obtained during an adsorption study describe the performance of the adsorbent and yield valuable information if properly interpreted. Several mathematical relationships have been developed to describe the equilibrium of solute between the solid and liquid phases and thus aid in the interpretation of adsorption data. These relationships apply when the adsorption tests are conducted at constant temperature and are referred to as adsorption isotherms, the three of the most common are Langmuir isotherm, the Freundlich isotherms, and the Brunaur Emmett-Teller (BET) isotherm.

3.2.2.1 Langmuir Isotherm Equation

Langmuir adsorption is a mathematical representation showing relationship between the amount of material adsorbed and concentration of material remaining in solution. It is based on the assumption that (1) a fixed number of accessible sites are available on the adsorbent surface, all of which have the same energy and adsorbent surface, all which have the same energy and that (2) adsorption is reversible. Equilibrium is reached when the rate to adsorption of molecules onto the surface is the same as the rate to desorption of molecules from the surface. The rate at which adsorption proceeds, then is proportional to the driving force, which is the difference between the amount adsorbed at particular concentration. Langmuir can be represented by the following mathematical relationship.

$$\frac{x}{m} = \frac{abc}{1+ac} \quad 3.1$$

$$\frac{1}{x/m} = \frac{1}{b} + \frac{1}{abc} \quad 3.2$$

where,

x = Amount of material adsorbed (mg)

m = Weight of adsorbent (mg)

c = Concentration of material remaining in solution after adsorption is complete (mg/1)

a and b = Empirical constants

If adsorption follows the Langmuir isotherm, a linear plot should result when the quantity 1/(x/m) is plotted against 1/C. Values of the empirical constant a and b can be determined from the slope and intercept of the plot.

3.2.2.2 Freundlich Isotherm Equation

Freundlich developed an empirical equation to describe the adsorption process. This is based on the assumption that the adsorbent had a heterogeneous surface composed of different classes of adsorption sites, with adsorption an each class of site following the Langmuir isotherm. Freundlich isotherm is given by the following equation.

$$\frac{x}{m} = KC^{1/n} \quad 3.3$$

$$\text{or, } \log\left(\frac{x}{m}\right) = \log K + \frac{1}{n}\log C \quad 3.4$$

where,

x = Amount of solute adsorbed (mg)

m = Weight of the adsorbent (mg)

C = Concentration of solute remaining in solution after adsorption is complete (mg/1)

K and n = Empirical constants.

Thus, a plot of log (x/m) verses log C, should yield a straight line for adsorption data which follow the Freundlich equation. Values of the constant n and K can be determined from the plot.

3.2.2.3 BET Isotherm Equation

Brunauer, Emmett, and Teller derived an adsorption isotherm based on the assumption that molecules could be adsorbed more than one layer thick on the surface of the adsorbent. Their equation, like the Langmuir equation, assumes that the adsorbent surface is composed of uniform, localized sites and that adsorption at one site does not affect adsorption at neighboring sites. Moreover, it was assumed that the energy of adsorption holds the first monolayer but that the condensation energy of the adsorbate is responsible for adsorption of successively layers. BET isotherm can be written as

$$\frac{x}{m} = \frac{AC \quad x_m}{(C_s - C)\left[1 + (A-1)\left(C/C_s\right)\right]} \qquad 3.5$$

$$\text{or, } \frac{C}{(C_s - C)\left(x/m\right)} = \frac{1}{A(X_m)} + \frac{A-1}{A(x_m)}\left(C/C_s\right) \qquad 3.6$$

where,

x = Amount of solute adsorbed (mg)

m = Weight of adsorbent (mg)

x_m = Amount of solute adsorbed in forming a complete monolyaer (mg/g)

C_s = Saturation concentration of solute (mg/1)

A = A constant to describe the energy of interaction between the solute and the adsorbent surface

Data from adsorption processes that conform to the BET equation will yield a straight line when left hand side of equation 3.6 is plotted against C/C_s)

3.3 Interpretation of Isotherms

Some typical isotherm shapes are shown as arithmetic graphs in Figure 3.1. qe is the amount of Cr (VI) adsorbed per unit mass of adsorbent is plotted on y axis and C is the equilibrium concentration of the aqueous solution. The linear isotherm goes through the origin, and the amount adsorbed is proportional to the concentration in the fluid. Isotherms that are convex are called favorable, because a relatively high solid loading can be obtained at low concentration in the fluid. The limiting case of a very favorable isotherm is irreversible adsorption, where the amount adsorbed is independent of concentration down to very low values. All systems show a decrease in the amount adsorbed with an increase in temperature, and of course adsorbate can be removed by raising the temperature even for the cases labeled irreversible. However, desorption requires a much higher temperature when the adsorption is strongly favorable or irreversible than when the isotherms are linear. An isotherm that is concave upward is called unfavorable because relatively low solid loading are obtained and because it leads to quite long mass transfer zones in the bed. Isotherms of this shape are rare, but they are worth studying to help understand the regeneration process. If the adsorption isotherm is favorable, mass transfer from the solid back to the fluid phase has characteristics similar to those for adsorption with an unfavorable isotherm (McCabe, *et al.*, 1993).

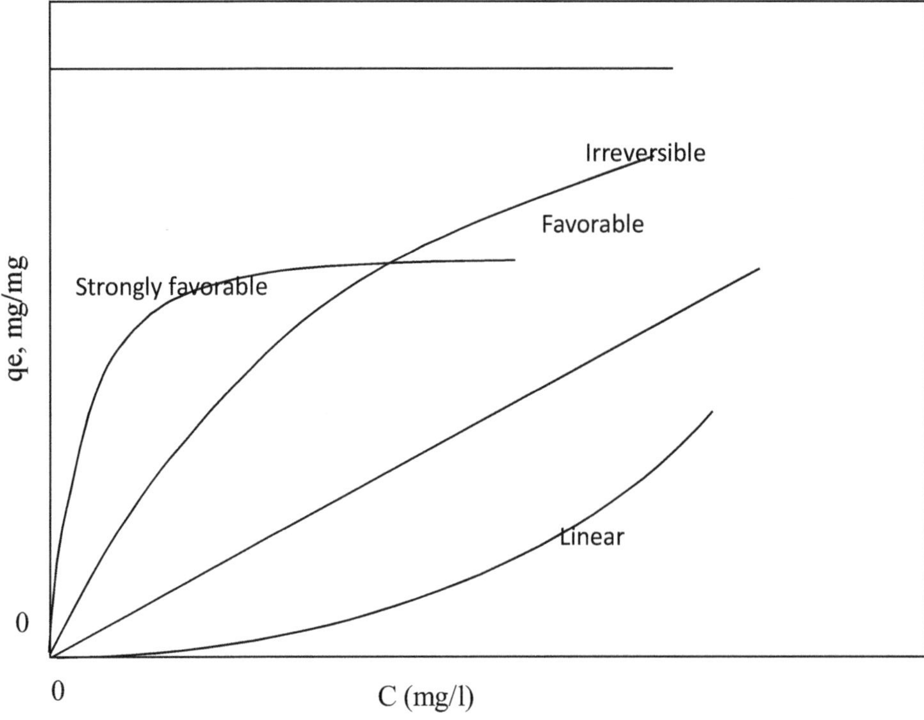

Figure 3.1: Adsorption Isotherms.

3.4 Types of Adsorption Systems

Adsorption process can be operated on either a batch or continuos flow basis.

3.4.1 Batch Adsorption Systems

In this system the adsorbent and wastewater are mixed together in a suitable reaction vessel until the concentration of solute has been reduced to the desired level. The number and size of reaction vessels and the amount of adsorbent required to treat the wastewater must be determined for the design of batch adsorption systems. These factors are influenced by wastewater volume, flow rate, and the rate of the adsorption reaction. Most batch adsorption systems are operated to small-scale wastewater volumes.

3.4.2 Column Adsorption Systems

This system is capable of treating large volumes of wastewater and are widely used for both municipal and industrial applications. Fixed bed adsorbers may be operated as single column or as multiple columns in series. They may be operated in either up flow or down flow mode. In down flow systems the adsorbent can serve for adsorption and for filtration of suspended solids; however, adsorption is more efficient and the beds are less expensive to operate if suspended solids are removed in advance. Up flow columns may operate either as packed or expanded beds. Packed beds require a high clarity influent to prevent clogging, whereas expanded beds are capable of handling wastewater high in suspended solids, since the solids will move through the void spaces between the adsorbent particle and not clog the bed. The use of up flow and down flow columns in series has been reported to optimize adsorbent usage and reduce operating costs.

3.5 Review of Recent Investigations

Naik, (2002) studied the environmental management including waste minimization for electroplating industry. The total water consumption during various processes were ranges from 1,000 lit/day up to 10,000 lit/day in typical electroplating shop in India. Various treatment options such as cyanide treatment, chromium treatment and final treatment are carried out in their study. Alkaline chlorination process was adopted in cyanide treatment in which chlorine was added in effluent containing free cyanides, in presence of sufficient alkali, the cyanides were oxidized first to cyanates and then to CO_2 and N_2. While hexavalent chromium was reduced to trivalent stage with the help of reducing agent such as sodium bisulphate in presence of sulfuric acid and then precipitated as trivalent $Cr(OH)_3$ with the help of alkali like sodium hydroxide or lime. The final treatment was done after reduction and removal of CN and Cr then the effluents are pumped into neutralization/settling tank. The effluents were neutralized and the salts are allowed to precipitate. The precipitates are taken to sludge drying beds and the clear effluent along the clear supernatant from the settling tank was neutralized and then discharged into sump from where it can be pumped to the disposal point.

Rao, *et al.* (2002) reviewed the various publications for the utilization of low cost adsorbents for the removal of heavy metals from wastewater. Author concluded in their study that most of the papers were based on the batch process. Little efforts have made on continuous and pilot plant scale studies, and recommended use of low cost adsorbents in wastewater treatment because of cheaper, locally available, required only simple alkali/acid treatment to increase their efficiency and can be regenerated for reuse with simple alkali/acid treatment or burned after drying.

Vasanth, *et al.* (2002) conducted experiments in batch process for the removal of colour malachite green onto carbonized paddy husk (CPH) using adsorption technique. About 90 per cent colour removal was obtained for a lower concentration of dye solution (50 mg/l). Effects of initial concentration, adsorbent dosage, and contact time, pH and agitation speed on the colour removal were studied. Rate of uptake of dye increased with decrease in initial concentration. Higher pH and adsorbent dosage increased rate of adsorption. Kinetics of removal has been found to follow the first order rate expression. The adsorption equilibria data fitted both Freundlich and Langmuir isotherms equally well. Particle diffusion studied indicates that predominantly adsorption of dye components take place rapidly followed by intraparticle diffusion.

De Castro *et al.* (2001) evaluated the sorption of heavy metals on a crude diatomite which was impregnated with a microemulsion which showed remarkable increase in chromium sorption capacity as compared to untreated diatomite. Samples with two different granulometries were investigated, both yielding practically complete adsorption. The adsorption process is pH dependent and the best results for the initial Cr (III) concentration of 1.5g/l were obtained at pH 2.95. The effect of the concentration of the chromium synthetic solution was also investigated. The adsorption isotherms were obtained (30, 40 and 50°C) and the Freundlich and Langmuir models were used to determine the adsorption capacity of the adsorbent. Following the adsorption step, a desorption process was carried out using several eluant solutions. The best desorption results were obtained using hydrochloric acid (100 per cent) as eluant.

Process development for the removal of lead and chromium from aqueous solutions using red mud an aluminium industry waste was investigated by (Gupta, *et al.*, 2001). The Red mud was converted in to an inexpensive and efficient adsorbent and used for the removal of lead and chromium from aqueous solutions. Effect of various factors on the removal of these metal ions from water (*e.g.* pH, adsorbent dose, adsorbate concentration, temperature, particle size, etc.) were studied. The effect of presence of other metal ions/surfactants on the removal of Pb^{2+} and Cr (VI) has also been studied. The material exhibits good adsorption capacity and the data follow both Freundlich and Langmuir models. Thermodynamic parameters indicate the feasibility of the process. Kinetic studies have been performed to understand the mechanism of adsorption. Dynamic modeling of lead and chromium removal on red mud has been undertaken and found to follow first-order kinetics. The rate constant and mass transfer coefficient have also been evaluated under optimum conditions of removal in order to understand the mechanism. Column studies have been carried out to compare these with batch capacities. The recovery of Pb^{2+} and Cr^{6+} and chemical regeneration of the spent column have also been tried.

Navarro, *et al.* (2001) have shown the influence of anions on the equilibrium and kinetic uptake of heavy metals from an aqueous solution by a novel nitrogen-type chelating adsorbent. Equilibrium experiments revealed that stoichiometric amount of metals and anions are adsorbed by the resin. Kinetic studies showed that during the initial stage of adsorption, the anions are adsorbed by the adsorbent prior to the metal ions. This occurred almost simultaneously with an increase in solution pH. At equilibrium, the pH returned towards its initial value. The concentration of anion also fluctuated during the entire equilibration process. Following these observations, mechanisms governing the role of anions to enhancing capacity and rate of metal uptake of this type of chelating adsorbent type were established.

The objective of the study (Ayub, *et al.*, 2001) was to evaluate the efficiency of neem bark in the treatment of industrial waste such as electroplating wastewater. The effects of pH, Contact time, Adsorbent dose, Concentration of metal, and Isotherm models were studied in a batch process. The removal was in general most effective at pH 3. At 50mg/l concentration of metal, about 83 per cent removal was observed in an adsorbent dose of 10g/l in 12 hours-contact time. The data for the above are fit well to the Freundlich isotherm.

Yoshio *et al.* (2001) have studied a recovery system of hexavalent chromium utilizing condensed tannin gels derived from a natural polymer with many polyhydroxyphenyl groups. The possible adsorption mechanism of Cr (VI) to the tannin molecules has been explained. The adsorption mechanism consist of four reaction steps; the esterfication of chromate with tannin molecules, the reduction of Cr (VI) to trivalent chromium Cr (III), the formation of carboxyl group by the oxidation of tannin molecules and ion exchange of the reduced Cr (III) with the carboxyl and hydroxyl groups. It was found in this recovery system that a large amount of protons were consumed accompanied with the reduction of Cr (VI) so that the acidic solution containing Cr (VI) was transferred automatically to neutral one by choosing an appropriate initial pH. The carboxyl group which was created by the oxidation of tannin molecule parallel to the reduction of Cr (VI) to Cr (III) contributed to an increase in the ion exchange sites of the reduced Cr (III). The maximum adsorption capacity of Cr (VI) reached 287 mg Cr/g dry tannin gel under the conditions of 0.77 water content of tannin gel and the initial pH of 2. This adsorption capacity was five to ten times higher than that obtained by the ion exchange between ordinary Cr (III) and tannin molecules for the tannin gels prepared under similar conditions. The proposed system had provided important information on a zero- emission- oriented process because it has such advantages as higher adsorption capacity of chromium and lower volume of secondary wastes compared with conventional process.

The ability of fruit peel of orange to remove Zn, Ni, Cu, Pb and Cr from aqueous solution by adsorption was studied by (Ajmal, *et al.*, 2000). It was found the adsorption was in the order of Ni (II) > Cu (II)> Pb (II) > Zn (II) > Cr (II). The extent of removal of Ni (II) was found to be dependent on sorbent dose, initial concentration, pH and adsorption followed first order kinetics. The process is endothermic showing monolayer adsorption of Ni (II), with a maximum adsorption to an initial concentration of 50 mg/l at pH 6.0. Thermodynamic parameters were also evaluated. Desorption was possible with 0.05 M HCl a 98.83 per cent in column and 76 per cent in batch process, receptively. The spent adsorbent was regenerated and recycled thrice. The removal and recycle in wastewater and was found to be 89 per cent and 93.33 per cent, respectively.

Peat has been investigated by several researchers as a sorbent for the removed of dissolved metals from waste streams. Besides being plentiful and inexpensive, peat possesses several characteristics that make it an effective media for the removal of dissolved metal pollutants. The mechanism of metal ion binding to peat has been explained with the several theories such as ion exchange, complexation, and surface adsorption being the prevalent theories. Factors affecting adsorption include pH, loading rates, and the presence of competing metals. The optimum pH range for metals capture is generally 3.5-6.5. Although the presence of more than one metal in a solution creates competition for sorption sites and less of a particular ion may be bound, the total sorption capacity has been found to increase. Studies have also shown that metals removal is most efficient when the loading rates are low. In addition, recovery of metals and regeneration of the peat is possible using acid elution with little effect on peat's sorption capacity. The utilization of peat and other biomass materials for the treatment of wastewater containing heavy metals is gaining more attention as a simple, effective and economical means of pollution remediation.

Pelleting processes can produce a robust media for a variety of applications where traditional methods of pollutant removal would be economically or technologically difficult (Brown, *et al.*, 2000).

Cimino, *et al.* (2000) showed the removal of ions such as Cd^{2+}, Zn^{2+}, three- and hexavalent chromium from aqueous solutions using hazelnut shell as biosorbent substrate. Batch equilibrium tests showed that the metal sorption was dependent on both pH and surface loading. For Cd^{2+}, Zn^{2+}, and Cr^{3+} ions the maximum removal was observed only into a specific pH range. The metal ion sorption obeyed both the Langmuir and Freundlich isotherms. Experiments by mixed solutions showed that more Cr^{3+} ions were removed than both $Cd^{2+}$and Zn^{2+} ions. The Cr (VI) removal was pH dependent and fitted with the Langmuir isotherm model.

Devies and Cliffe, (2000) investigated the removal of Zn and Cu from sea water using mucas of the littoral gas tropod mollusc patella vulgata (common limpet), which were collected from the coast of north east England and used within 36 hours of collection. Authors reported the detection limits were 0.01µg/g for Zn and 0.03µg/g for Cu.

Nakano, *et al.* (2000) developed tannin gel particles those had extremely high adsorption capacity for hexavalent chromium Cr (VI) for controlling gelation of Mimosa tannin extracted from the bark of plants. The gelation process consisted of two stages; partial gelation of Mimosa tannin by reaction with formalde- hyde (cross-lining agent) and granulation by dispersing the partially gelated solution into a liquid mixture of decalin and polyether nonionic surfactant with vigorous stirring. The maximum adsorption capacity, 540 mg-Cr (VI)/(g-dry tannin gel) was obtained at a water content of 77.6 per cent with an acidic solution of pH 6.2.

For removal of chromium using low cost adsorbents, Rao, *et al.* (2000) have applied adsorption technique for the removal of chromium from aqueous solution using wheat straw dust, and coconut jute, and compared the results with the powdered activated carbon. The high uptake of hexavalent chromium was observed with powder activated carbon at pH 2.0, and for the other adsorbents at pH 6.0. The sorption of Cr (VI) is found to be diffusion controlled and the external mass transfer were determined by varying contact time, adsorbent dose, initial Cr (VI) concentration and the adsorbent particle size at a temperature of $38 \pm 1°$ C. Batch adsorption studies were carried out and the Langmuir and Freundlich adsorption isotherms were used to analyze the equilibrium data. Powdered activated carbon had removed 90 percent of Cr (VI) at pH 2.0, for an initial concentration of 50mg/l, adsorbent dose of 2.5 g/l and contact period of 60 minutes, where as the wheat straw dust and coconut jute effectively removed the hexavalent chromium at pH 6.0 with an adsorbent dose of 3.0 mg/l, initial concentration of 100 mg/l, contact time of 120 minutes and 150 minutes for coconut jute. The Langmuir isotherm is the better fit with respect to Freundlich isotherm for all the adsorbents.

Neem (*Azadirachta indica*) bark for the removal of mercury from water (Ansari, *et al.*, 1999) was studied for the removal of mercury from the water. They observed that neem bark powder of <53µ size has very high sorption capacity for Hg (II) with

> 99 per cent removal from the solution containing 100mg/l of Hg (II) with 0.1 per cent (W/V) concentration of neem bark powder.

Aggarwal, *et al.* (1999) studied the adsorption isotherms of Cr (III) and Cr (VI) ions on two samples of activated carbon fibres and two samples of granulated activated carbon solutions in the concentration range 20-1000 mg/l. The adsorption isotherms were determined after modifying the surfaces by oxidation with nitric acid, ammonium persulphate, hydrogen peroxide and oxygen gas at 350° C and after degassing at different adsorption of Cr (II) ions increases on oxidation and decreases on degassing. On the other hand, the adsorption of Cr (VI) ions decreased on increase of degassing. The increase of Cr (II) and the decrease of Cr (VI) on oxidation and decrease of Cr (III) and increase of Cr (VI) have been attributed to the fact that the oxidation of the carbon surface enhances the amount of acidic carbon - oxygen surface groups while eliminates these groups. Thus while the presence of acidic surface groups enhance the adsorption of Cr (II) cations, it suppresses Cr (VI) anions.

Ariyapadi, *et al.* (1999) found the high degree of Cr (VI) removal using simple cheletan type ion exchange resin. A synthetic effluent containing 200 mg/l of Cr (VI) was allowed to pass through the column at a flow rate of 2.5 l/h. Results showed that Cr (VI) concentration in the outlet remained less than 0.1 mg/l for about 5 bed volumes and then as the bed saturated it began to rise sharply and then the concentration of Cr (VI) remained constant. The resin was regenerated using 5 per cent HCl. Investigators concluded that the structure of the resin could not be clearly understood, hence future researchers may work on the maximum number of cycles and the bed volumes, effect of impurities and interfering agents on the resin and need to develop a suitable model for the adsorption mechanism.

The use of low-cost sorbents has been investigated as a replacement for current expensive methods of removing heavy metals from solution (Bailey, *et al.*, 1999). Natural materials or waste products from some industries which show a high capacity for heavy metals removal can be used, and disposed off safely with little cost. Modification of the sorbents can also improve adsorption capacity. Bailey *et al.* (1999) have given an extensive list of sorbent and a summary of available information on a wide range of potentially low-cost sorbents, including bark, chitosan, xanthate, zeolite, clay, peat moss, seaweed, dead biomass, and others has been given. Some of the highest adsorption capacities reported for cadmium, chromium, lead and mercury are: 1587mg Pb/g lignin, 796mg Pb/g chitosan, 1123mg Hg/g chitosan, 1000mg Hg/g CPEI cotton, 92mg Cr (III)/g chitosan, 76mg Cr (III)/g peat, 558mg Cd/g chitosan, and 215mg Cd/g seaweed.

Bhalke, *et al.* (1999) studied the uptake of heavy metals including fertile uranium and fission products strontium and cesium by using sunflower plant dry powder at pH 2.5, 4.0 and 7.5. The heavy metals were taken up by the plant to significant levels (K_d> 10 mg/l at pH 7.5). The distribution coefficient (K_d) value for Sr is higher at pH 2.5 than at pH 4.0 and 7.5. Other elements such as Pb, Zn, Cu, Cd have lowest K_d values at pH 2.5 and highest at 7.5 in the descending order.

Contrary to using natural adsorbents, some investigators worked with the synthesized reactive polymer such as, long alkyl quaternized poly (4-vinylpyridine) (PVP), and coating it on the surface of silica gel to produce a granular sorbent to remove Cr (VI) from water. Batch experiments were conducted to determine the kinetics, sorption isotherm, pH effects, and influence of other anions on the chromium adsorption onto the coated silica gel. The research demonstrated that the synthesized PVP-coated silica gel (referred to as coated gel) could successfully remove chromium (VI) from solution. The adsorption of Cr (VI) by the coated gel was strongly influenced by the pH. The maximum sorption occurred at about pH 4.5-5.5 under the laboratory conditions. The removal efficiency was 100 per cent when the initial Cr (VI) concentration was 2.5 mg/l with 2.5 g/l of coated gel at pH 5.0. The concentration of Cr (VI) had pronounced effect on the rate of sorption. Compared to ion exchange, the sorption kinetics of Cr (VI) was rapid (about 5 hrs.) The equilibrium sorption data fitted the Langmuir isotherm model. Chromium adsorbed on the coated gel was easily recovered under certain conditions (Gang, *et al.*, 1999).

Oat and wheat straw are abundant yet at the same time are not widely used. For this reason (Rios, *et al.*, 1999) tested them as metal ion removers from water solutions. Batch experiments were done to determine the affinity of both biomasses for Cu (II), Pb (II, Cr (III), Cr (VI), Ni (II), Zn (II), and Cd (II) metal ions at several pHs; times of reaction and the adsorption capacity for each metal were determined under optimum conditions. The metal ion removal from the biomass was also measured. The batch experiments showed that pH 5 was the best for adsorption of most of the metal ions, the exception being Cr (VI) (pH 2) in both biomasses. The best adsorption capacities were observed for Pb (II), and Cd (II) for both biomasses. The biomasses were also tested after modification with NaOH to improve their adsorption capacities. The NaOH modification conditions were established by experiments carried out on each biomass. Surprisingly, it was noticed that the drying process during modification affects the adsorption capacity of oat biomass but not wheat biomass. Simultaneous experiments for adsorption capacities of each metal ion were performed by exposing both the unmodified and NaOH modified biomasses to the same environmental conditions. It seems that both biomasses can be employed in heavy metal ion removal processes.

Environmental pollution caused by chromium is of considerable importance. It finds wide spread usage in electroplating, metal finishing, leather tanning industries. Chromium compounds are toxic substances that have adverse effects on living organism. Hence the removal of chromium from wastewater prior to its disposal is essential. An attempt has been made in the study to evaluate the potential of lignite to remove hexavalent chromium from the synthetic wastewater by adsorption. In order to understand the adsorption behavior and the adsorption potential of lignite, a number of batch experiments were conducted by (Ayub, *et al.*, 1999). The effects of initial Cr (VI) concentration, contact time, and adsorbent dosage on the potential of lignite to remove chromium from wastewater were studied. The results obtained indicate the feasibility of lignite to be used as an adsorbent for the removal of chromium from the wastewater. The adsorption behavior followed the Freundlich isotherm.

In the study of (Tran, *et al.*, 1999), desiccant silica gel (DSG) was compared with chromatography silica gel (CSG) for its ability to remove metal ions including Pb^{2+}, Cu^{2+}, Ni^{2+}, Zn^{2+}, Cd^{2+} and UO_2^{2+} from solution. The equilibration time was shorter for UO_2^{2+} (less then 1 h) than for the heavy metal ions (2-3h) and adsorption by DSG took longer to reach equilibrium compared with CSG, probably due to the smaller mean diameter of its pores. The adsorption process showed first order kinetics for all the metals studied. The mass transfer coefficients and overall rate constants were determined for each of the metals. The adsorption rates of the metal ions, in order of decreasing magnitude, were $UO_2^{2+}>Pb^{2+}>Cu^{2+}>Zn^{2+}>Ni^{2+}>Cd^{2+}$ for both DSG and CSG. Metal uptake was found to increase rapidly within narrow pH range characteristic for each metal. The adsorption of the metal ions obeyed the Langmuir isotherm and followed the preferential order $UO_2^{2+}>Pb^{2+}>Cu^{2+}>Zn^{2+}>Ni^{2+}>Cd^{2+}$. Similarity of the isotherms for the individual metals was observed for both DSG and CSG, indicating a comparable removal capacity of DSG and its possible use as an alternative to CSG in adsorption applications. Adsorption from lead (II)-uranium (VI) solutions showed that the uptake of each metal was considerable reduced with an increasing concentration of the other, the adsorption of lead (II) being more strongly influenced by uranium (VI) than vice versa due to the higher affinity of silica gel for the latter.

Ajmal, *et al.* (1998) studied that adsorption behavior of cadmium, zinc, nickel, and lead from aqueous solutions by manngifera idica seed shell. Authors conducted the batch adsorption studied at different temperatures. Adsorption increased with contact time and equilibrium attained in 90 minutes, it was observed that 72 per cent of cadmium removal at pH 5.0 and 91 per cent removal of Zn, Ni and Pb at 6.0 and thereafter increase in pH resulted in decrease in percent removal. In case of Cd (II), the maximum removal was found when the temperature was between 30- 40° C, but above 40° C the adsorption of the adsorbate increase may be due to weakening of adsorptive forces between active sites of the adsorbents and the adsorbate species and also between the adjacent molecules of the adsorbed phase. Adsorption of Cd (II), Zn (II) and Pb (II) at pH 5.0 obeyed the Freundlich equation and the Langmuir equation. The calculated Freundlich equation constants (ln K) ware > 1, at all the temperature range (40- 50° C), indicated that the ions Zn, Ni and Pb are strongly adsorbed on the adsorbent. Finally authors studied the effect of salinity by adding NaCl (0.25-9.0 gm/50 ml) on the adsorption of Cd, Zn, Ni and Pb. It was found that the presence of NaCl reduces the adsorption in all cases, may be due to relative competition between Na ions and Cd, Zn, Ni and Pb species on the active adsorption sites of Mangifera India seed shell.

Bandyopadhyay, *et al.* (1998) made an attempt to investigate the adsorption of hexavalent chromium on sand alone and in combination with activated carbon. Authors reported that mixed adsorption (namely mixture of sand and activated carbon) brought out some interesting observations on the theory of adsorption. The removal of hexavalent chromium was found to be synergistically modified by the presence of sand over the activated carbon. Results indicated that equilibrium was attained in 2 hrs. For Cr (VI) removal by activated carbon the maximum removal of 72 per cent was achieved at an adsorbent dose of 2g/100 ml for an initial Cr (VI) concentration of 2 mg/l. Optimum removal has been found to occur at a mixed adsorbent dose composition of 70 per cent activated carbon and 30 per cent sand by mass.

De Arnab Kumar, (1998) reported in his review paper that the percentage removal of various heavy metals such as Hg, Pb, Cd, Cr, Fe, Mn, Cu, ans Ni using fly ash as low cost adsorbent are 93 per cent, 92 per cent, 93 per cent, 44 per cent, 99 per cent, 58 per cent, 95 per cent, and 99.5 per cent respectively. They reported the removal of Hg from the waste containing Hg (II) and concentration was reduced from 47 ppm to 0.76 ppm with an adsorbent (Coal fly ash) dose of 4.42 ppm. If the initial concentration of adsorbate is more, then higher bed depth was required. The exhausted fly ash was regenerated using 0.1N HCL or 0.1 N H_2SO_4.

Raji, *et al.* (1998) used saw dust charcoal for the removal of hexavalent chromium. Batch experiments were carried out and observed that at pH 2.5 and temperature 30° C, the equilibrium for the removal of Cr (VI) reached within 210, 240, 270 and 300 minute for an initial concentration of 30, 75, 100 and 150 mg/l respectively. With the increase in initial concentration of Cr (VI) removal increased from 86.1 per cent to 95.2 per cent at pH 2.5 and 30° C. This is due to that the percentage uptake is dependent on initial concentration because the adsorption sites adsorbed the available chromium more quickly at low solute concentrations. However, for higher concentrations, interparticle diffusion was predominant adsorption mechanism. The effect of pH on Cr (VI) removal was studied and found that the percentage removal increases to maximum at pH 2.5 and thereafter decreases sharply.

Saravanne, *et al.* (1998) studied the adsorption of Cu (II), Mn (II), Fe (II), and Cd (II) on saw dust, rice husk and chemically modified saw dust and the rice husk. Rice husk and saw dust was activated by treating three parts of its (by weight) with 1 to 2 parts of EDTA or activated carbon and keeping it in hot air oven at a temperature of 140 to 160 ° C for a period of 24 hours. It was washed well with water to remove the traces of free acid and then dried at 105 to 110 ° C and then activated at a higher temperature at 800 to 850 ° C for a period of 30 minutes. It was found that the saw dust posses greater adsorption capacity for all metals than rice husk. Further chemically modified saw dust could remove 98.28 per cent Cu (II), 100 per cent Mn (II), 96.72 per cent Fe (II) and 96.72 per cent Cd (II) from the wastewater. The chemically treated rice husks and saw dust both are very efficient for the removal of metal. It was found that the cupper and cadmium adsorption are effective at pH 4-6, manganese at pH 7 and the iron at pH 3-4. Removal of Cu (II), Mn (II) and Cd (II) increases from 28 mg/g to 48 mg/g when the concentration of metal ions increases from 30 to 100 mg/l for all metals over a pH range of 2 to 9. It was also observed that the equilibrium was attained within 2 to 3 hours and is independent of initial concentration. Authors concluded that the activated carbon as an activator is found to be superior to EDTA for saw dust and rice husk and the process of uptake obey both Langmuir and Freundlich isotherm.

Feasibility of foundry material, Wollastonite as adsorbent has been examined by (Sharma, *et al.*, 1998) for water pollution control using adsorption process. The removal is found to be concentration dependent and low concentrations favour the uptake. The uptake increased from 41.7 to 69.5 per cent by decreasing the concentration from 2.0×10^{-4} M to 0.5×10^{-4} M at 0.01 M $NaClO_4$ ionic strength, 2.5 pH and 30°C. Rate of uptake was found to be 3.0×10^{-2} min^{-1} under optimum conditions and the process is governed by first order kinetic equation. The process involves both film and pore diffusion, the min^{-2} under favourable conditions, while mass transfer coefficient for film mass transfer was found to be 2.31×10^{-3} s^{-1} at 0.5×10^{-4} M Cr (VI) concentration at optimum conditions. The process is a typical example of endothermic adsorption and therefore, higher temperatures favour the uptake. Various empirical models have been developed.

Ayub, *et al.* (1998) reviewed the earlier work for the removal of heavy metals using agricultural waste products. His survey reveals that stated that the various researchers have used low cost adsorbents such as saw, dust, straw, jute etc., in their studies. Although these adsorbents are cheaper then activated carbon but they are not feasible for industrial usage. These adsorbents usually clog the treatment system and their volume increases when they come in contact with moisture.

The adsorption of nitrate, chromium (VI), arsenic (V) and selenium (VI) anions in an amine modified coconut coir (MCC- AE : with secondary and tertiary amine functionality) was studied by (Baes, *et al.*, 1997) to determine the capability of easily prepared and low cost materials in removing typical groundwater anions contaminants. Batch adsorption ion exchange experiments were conducted using 200 mg MCC- AE, initially containing chloride as the resident anion, and 50 ml of different anion containing water of varying concentrations. It was presumed, at low pH only SeO_4^{2-} remained as divalent anion, while monovalent species $H_2AsO_4^-$ and $HCrO_4^-$ predominated in their respective exchanging solutions. The adsorption data were fitted into the Freundlich Model and maximum adsorption for each anion was estimated using their respective Freundlich equation constants.

Gupta, *et al.* (1997) used activated carbon developed from fertilizer waste for the removal of Hg (II), Cr (VI), Pb (II) and Cu (II). The raw material was a waste product collected from National Fertilizer Limited (NFL) and was converted into activated carbon and the particles in the size range of 200 to 250 mesh were selected. Parameters selected for the studies were length of the primary adsorption zone (PAZ), total time involved for PAZ, time for PAZ to move down its length (t_d), amount of adsorbate adsorbed in PAZ from break point to exhaustion, time of initial formation of PAZ (t_f), etc. From the experimental results authors concluded that the total time for PAZ to establish itself move down the length of the column and the out of bed is least for Hg (II) and maximum for Cu (II) while Pb (II) and Cr (VI) falls in between. Similar trend was found for t_d and t_f. The time required for the movement of zone down of its own length in the column is that in the between 2 to 4 hours. The time required for the formation of initial PAZ is between 1 to 2 hours. The percent saturation at break point is 79.2, 75.0, 66.7 and 64.3 per cent for Hg^{+2} Pb^{+2} Cr^{+6} and Cu^{+2} respectively. Exhausted adsorbent can be regenerated by applying 50.0 ml of NH_4OH for almost complete desorption of chromium and 80.0 ml of 2M HNO_3 for lead. Authors have considered the cost of the adsorbent and found that the most inexpensive variety of commercially available carbon in India costs US$250/ton where as fertilizer waste costs about US$20/ton including the cost of conversion into activated carbon.

Gupta, *et al.* (1997) studied the blast furnace slag, a waste generated in steel plant and converted in to a low cost potential adsorbent for the removal of Zn and Cd from wastewater. The optimum conditions were found to be pH 6.0 for Zn and 5.0 for Cd. The uptake of the metal ions was 75 - 90 per cent at the low concentrations. Also the uptake of Zn and Cd increased with an increase in temperature thereby indicating the process to be endothermic in nature. The concentration of adsorption sites may be increased with raising temperature due to breaking of some internal bonds near the edge of particle. Adsorption data were fitted to Freundlich and Langmuir isotherms and the adsorption capacity (K_f) is less for zinc slag system than that for the cadmium slag system. The value of Langmuir constant appeared to be significantly higher for the zinc slag system in comparison to the uptake of cadmium on the same adsorbent. Column operations were also performed and used BDST model to fit the data.

Fe (III), Cr (III) hydroxide, a waste by - product obtained from the treatment of Cr (VI) containing wastewater in a fertilizer industry has been used by (Namasivayan, *et al.*, 1997) for the adsorption of Hg (II) from the aqueous solution. The influence of various parameters such as metal in the concentration (10-40-mg/l), agitation time (1-100 minutes), adsorbent dose (5-250 mg/50 ml), temperature (24 - 44° C) and pH value (4-10) on the removal of Hg (II) were studied. The adsorption followed both Langmuir and Freundlich isotherm models. The applicability of Lagergren model was also investigated. Almost complete removal of Hg (II) from the solution containing 40 mg/l in 50 ml solution by 175 mg of adsorbent occurred at initial pH of 5.6. Adsorption was uniformly high (91 per cent) in the pH range of 4.0 to 10.0. Desorption of Hg (II) showed that it was solubilised in 2 per cent KI to the extant of 65 per cent.

Carbon slurry, generated as a waste material in a naphtha-based ammonia plant of the Fertilizer Corporation of India, Gorakhpur, has been used by (Singh, *et al.*, 1997) as an adsorbent for the removal of Cr (VI) from aqueous solution at different experimental conditions. The removal was favoured at low pH, with maximum removal at pH 2.5. The effects of concentration and temperature have also been reported. Batch adsorption kinetics have been described by the Lagergren equation. The applicability of the Langmuir isotherm for the present system has been tested at different temperatures. Thermodynamic parameters indicate the endothermic nature of Cr (VI) adsorption on carbon slurry. Recovery of adsorbed chromium for reuse has also been reported in the study.

Sharma, *et al.* (1996) studied the foundry material for the removal of Cr (VI). It was observed that the removal increased from 41.7 to 69.5 per cent by decreasing the concentration of Cr (VI) 2.0×10^{-4} M to 0.50×10^{-4} M at 0.01 M $NaClO_4$ ionic strength, pH 2.5, temperature 30° C and particle size of 100 micrometer.

Srivastava, *et al.* (1996) studied that the removal of lead and chromium from wastewater using activated carbon developed from fertilizer waste material. Authors found that the removal of Cr (VI) was maximum at pH 2.0, where as in case of lead it did not change much with increase in pH and 50-80 per cent of the removal occurred within the first hour of contact time. The optimum parameters were found as dose of adsorbent as 1.0 g/l, temperature 30 ° C, particle size of 150-200 mesh and adsorbate concentration of $1x10^{-2}$M.

Patnaik, *et al.* (1995) used blast furnace flue dust generated in the steel plants for the removal of hexavalent chromium. Results showed that at lower initial concentration of Cr (VI), the efficiency of removal of Cr (VI) is higher at lower pH. Similarly there was a definite decrease in percent removal from 16.3 to 10.1 per cent with the increasing temperature from 301 to 323 k. With the increase in adsorbent dose from 10 to 15 mg/l, Cr (VI) removal efficiency increases to more than 90 per cent when the Cr (VI) concentration is in between 2 to 5 mg/l.

Sujata, *et al.* (1995) used the spent myrobalan nuts, a waste product for the removal of Cr (VI). It was found that 1.0 g/l of spent nut removes 97 per cent of Cr (VI) at optimum contact time of 75 minutes and pH of 2.5, for the concentration of 5.0 mg/l.

Chand, *et al.* (1994) carried out batch experiment to assess the suitability of bagasse and coconut jute based materials as adsorbents for the Cr (VI) removal. The effects of solution pH, Cr (VI) concentration, adsorbent dosage and contact time were studied. The removal was in general most effective at low pH (≤ 2) values and low chromium (VI) concentrations. Activated coconut jute carbon was the most active among the four adsorbents studied. It was fairly stable even at higher pH. Removal of Cr (VI) to the extant of 97 per cent, was observed with coconut jute carbon at natural pH, whereas other conventional adsorbents showed much lower activities. The characteristics of the adsorbents (Chand, *et al.*, 1994 and Pillai, *et al.*, 1981) are shown in Table 3.1.

Table 3.1: Characteristics of the Adsorbents

Absorbent	*Per cent Moisture*	*Per cent Volatile Matter*	*Per cent Ash*
Raw bagasse	8.2	82.9	8.9
Activated bagasse carbon	7.5	20.7	71.8
Activated coconut jute carbon	6.8	17.2	76.0
Bagasse ash	0.1	9.6	99.3

Composition of Coconut Shell (Pillai, et al., 1981) per cent dry nut

Moisture	*Ether Extract*	*Alcohol Benzene*	*Hot Water Extract*	*Pentosan*	*Lignin*	*Cellulose*
6.76	0.17	1.98	1.76	30.31	32.32	26.60

Singh, *et al.* (1994) investigated the removal of Cr (VI) leached acacia arabika bark which was treated with formaldehyde in acidic medium to prevent its colour leaching tendency. Sorption of Cr (VI) decreased with increase in pH (9.25 mg/g at pH 2.0 to 2.5mg/g at pH 6.8). The equilibrium time was found to be two hours. For complete removal of 50 mg/l of Cr (VI), adsorbent having a particle size of

0.44 and 1.22 mm was required in quantities of 3 and 10 mg/l respectively and the adsorption data followed the Langmuir isotherm.

The adsorption isotherm of chromium (VI) on activated carbon was obtained by (Ramos, *et al.*, 1994) in a batch adsorber. The experimental adsorption data were fitted reasonably well to the Freundlich isotherm. The effect of pH on the adsorption isotherm was investigated at pH values of 4, 6, 7, 8 and 12. It was found that at pH <6, Cr (VI) was adsorbed and reduced to Cr (III) by the catalytic action of the carbon and that at pH ≥ 12, Cr (VI) was not adsorbed on activated carbon. Maximum adsorption capacity was observed at pH 6 and the adsorption capacity was diminished about 17 times by increasing the pH from 6 to 10. The pH effect was attributed to the different complexes that Cr (VI) can form in aqueous solution. The adsorption isotherm was also affected by the temperature since the adsorption capacity increased by raising the temperature from 25 to 40°C. It was concluded that Cr (VI) was adsorbed significantly on activated carbon at pH 6 and that the adsorption capacity was greatly dependent upon pH.

Sharma, *et al.* (1993) examined the adsorption of hexavalent chromium from aqueous solution by Irish sphagnum moss peat, studied the thermodynamics and kinetics of adsorption (and desorption) and reported the effect of pH and temperature on the adsorption. Two important physico-chemical aspects for the evaluation of the sorption process are the equilibria of sorption and the kinetics. Sorption equilibrium is established when the concentration of metal in a bulk solution is in dynamic balance with that of the interface. It is generally seen that the adsorption density, which is the mass (mg) of total chromium, removed per unit mass (g) of peat, increases slowly with decreasing solution pH up to a value of 2.5 and then rather sharply with further decrease in pH up to 1.5. Isothermal data have been used to calculate the ultimate sorption capacity of the peat by substituting the required equilibrium concentrations in the Langmuir and Freundlich equations. The data demonstrate that peat is an effective adsorbent for chromium (VI) in acidic solutions in the 1.5-3.0 pH range.

The effect of pH on adsorption process of a new low cost adsorbent material 'rice straw' (ORYZA SATIVA) was studied by Ali, *et al.* (1992). The adsorption process of Cr (VI) on straw was found to be fully pH dependent at room temperature. The percent adsorption of Cr (VI) varies from 100 per cent at low pH of 1-3 to 60-70 per cent for solution of pH from 4 to 12 from its dilute aqueous solution. The process of adsorption of Cr (VI) on straw, though involves the pH adjustment but is effective, economical and convenient the adsorption is found to be complete. The adsorption of Cr (VI) on straw was also found to be time dependent and effective for removal of Cr (VI). For it both time and pH need to be considered.

The ability of a low cost adsorbent material in removal of heavy metal ion Cr (VI) from aqueous solution was successfully investigated by Deo, *et al.* (1992). The material, paddy straw was found to be effective and was capable of removing completely the Cr (VI) from dilute aqueous solution. However, at higher initial concentration of Cr (VI) the percent removal was low. The method is effective under normal condition and needs no adjustment of pH and is cheap and convenient. The used adsorbent could be disposed off easily by burning the material. The dynamics of adsorption from bulk to solid phase under various experimental conditions was studied, and the results were interpreted with the help of various related isotherms.

Singh, *et al.* (1992) used saw dust coated with the iron hexamine gel for the removal of certain heavy metals such as Hg (II), Pb (II), Cr (VI), Ni (II), Cd (II) and Cu (II). This modified adsorbent exhibited good adsorption potential for Hg (II), Pb (II) and Cr (VI) and significant uptake of Ni (II), Cd (II) and Cu (II) but had poor capacity for Zn (II) and Mn (II) removal.

Deepak, *et al.* (1991) used activated carbon and alumina for the removal of Cr (VI). It was found that 5g of activated charcoal could reduced 100 per cent Cr (VI) from a 200 ml sample containing 200 mg/l Cr (VI) at neutral pH. With increase in initial Cr (VI) concentration there was corresponding decrease in percent Cr (VI) removal, because of the limitations of surface area for adsorption. For the same concentration alumina (Al_2O_3) showed 58 per cent removal when 5.0 g/200 ml of adsorbent dose was given.

Vanangamudi, *et al.* (1991) reviewed the chromium removal technologies. Chromium compounds are toxic substances which have pronounced adverse effects on human beings, animals, plants and aquatic life. Hence, the removal of chromium from the wastewaters becomes important. Adsorption is highly effective, cheap and easy method for the removal of chromium from effluents. Activated carbon, which is frequently used in the adsorption of pollutants, is costly both to use and to regenerate. Therefore, there is need for development of low cost, easily available materials which can remove chromium efficiently from various wastewaters. In this paper an attempt has been made to review a variety of low cost adsorbents used by various researchers.

Manju, *et al.* (1990) investigated the removal of Cr (VI) using coconut fibre pith (CEP). Batch adsorption studies were conducted to determine the effect of adsorbent dose and adsorbate concentration. Maximum removal of hexavalent chromium was found in first two hours and equilibrium attained in three hours. Also it was found that percentage Cr (VI) removal decreased from 99.2 per cent to 87.39 per cent with an increase in initial concentration of Cr (VI) from 50 to 200 mg/l. It was reported that this percentage reduction might be due to the fact that for a fixed adsorbent dose, the total available sites are limited thereby adsorbing almost the same amount of chromium. Authors concluded that the process is exothermic and the maximum Cr (VI) removal occurs at pH 2.0 and the thermodynamic parameters have been calculated and the adsorption data fits the Freundlich isotherms equation. They also reported that the spent carbon could be regenerated by washing carbon with 1M NaoH.

Removal of Chromium by adsorption on bituminous coal from the synthetic effluent is reported by Krishnaiah, *et al.* (1989). In order to understand the sorption behavior of Chromium and also to evaluate the extent of reduction from Cr (VI) to Cr (III), number of batch experiments were conducted on each-adsorbent-adsorbate system at various pH values (pH 1-6) with different amounts of adsorbent. The results illustrate that the adsorption is maximum at pH 2 and reduction is maximum at pH 1.

Tiwari, *et al.* (1989) attempted to remove chromium (VI), copper and nickel from dilute solutions (concentrations ranging from 5 to 50 mg/1) by an activated carbon adsorption technique. Studies were restricted to pH range from 5.5 to 8.0. Adsorption of chromium (VI) was found to exhibit a peak value at pH 5.5 and copper at 8.0. The adsorption of nickel was found to be appreciable in both the mediums but removal was faster in alkaline medium.

Mehrotra, *et al.* (1988) used raw rice husk, raw bone powder, burnt rice husk, burnt bone powder and hair for the removal of chromium (VI) from aqueous solution. Results showed that major part of the total adsorption gets completed within 6-12 hours for all the adsorbents at all pH values and for hair and burnt rice husk the maximum uptake was obtained in 12 hrs at pH 2.0. The equilibrium data were approximated to the Freundlich and Langmuir isotherms for all the cases, hence it was a monolayer adsorption. Authors concluded that at an initial Cr (VI) concentration of 12,500 mg/l and 50,000 mg/l, Freundlich constant (1/n) was found to be > 1 and < 1 respectively indicating that adsorption was more efficient in lower concentrations. Authors also reported that all the adsorbents tried, hair and burnt rice husk exhibited good sorption capacity and hair was effective in removing various cations such as Hg^{2+}, Cu^{2+}, Cd^{2+}, Ni^{2+} and Pb^{2+} from aqueous solutions. Activated carbon exhibits constant adsorption capacity over a wide pH range.

Srinivasan, *et al.* (1988) reported that the removal of Cr (VI) by rice husk carbon and compared the results with commercial activated carbon. It was reported that 88 per cent removal of total chromium and 99 per cent removal of hexavalent chromium in the pH range of 2.0-3.0 with a carbon dose of 1.6 g/l and an equilibrium time of 4 hours.

Tare, *et al.* (1988) studied the use of soluble starch xanthates (SSX) for the removal of Cd (II), Cr (VI) and Cu (II) and insoluble starch xanthate (ISX) for the Cr (VI) and Cu (II). Results showed that ISX has better binding capacity for metals. However with due consideration to yield the chemical requirements for the syntheses of ISX and SSX, SSX appears to have higher adsorption capacity for metal removal. The order of preference of metal binding capacity of SSX was found to be as Cr (VI) > Cu (II)> Cd (II) whereas ISX as Cr (VI) > Cu (II).

Ananthakrishna, *et al.* (1982) investigated the removal of Cr (VI) by a batch process using the bottom ash as an adsorbent and achieved the maximum percentage removal (93 per cent) at pH 4.0 with an initial chromium (VI) concentration of 200 mg/l. Authors reported that as the Cr (VI) concentration increases the removal capacity also increases.

The removal of hexavalent chromium, Cr (VI) from dilute aqueous solution by activated carbon has been investigated by (Huang and Bowers, 1978). Hexavalent chromium species were removed by adsorption onto the carbon surface or by reduction to the trivalent state. The effects of hexavalent chromium concentration, carbon dosage, pH, and mixing on the rates of reduction and adsorption were studied in batch experiments.

Huang, *et al.* (1975) investigated the use of calcinated coke for removal of chromium from dilute aqueous solution by a continuously mixed batch system. It was found that the percentage of chromium removal is greatly affected by chromium present or more specifically, by the relative concentration of chromium and coke surface. At pH 2, for instance, the percentage for chromium removal increases from 7 percent for granular coke and 16 percent for powdered coke, respectively, at a hexavalent Cr concentration of 312 mg/1 as Cr to 100 percent for both cokes at 5.2 mg/1 as Cr. In general, more than 50 percent removal efficiency may be achieved if the hexavalent Cr concentration is less than 26 to 52 mg/1 as Cr and pH is less than 2 to 3.

Chapter 4

Objective of the Present Work and its Relationship with the Previous

Various investigators (Aggarwal, *et al.*, 1999; Deepak, *et al.*, 1991; Mahesh, *et al.*, 1999; Ramos, *et al.*, 1994; Singh, *et al.*, 1994; Srivastava, *et al.*, 1996; Tiwari, *et al.*, 1989; Huang, *et al.*, 1977; Alerts, *et al.*, 1989; Kim, *et al.*, 1977; Lee, *et al.*, 1989; Gang, *et al.*, 1999; Deo, *et al.*, 1992; Huang, *et al.*, 1975; Periasamy, *et al.*, 1991; Ayub, *et al.*, 1998, 1999, 2001, 2002; Drake, *et al.*, 1996; Shukhla and Sakhardane, 1991; Weber, 1996; Chand, *et al.*, 1994; Siddiqui, *et al.*, 1994; Camino, *et al.*, 2000) have studied the different techniques for the removal of heavy metals from wastewater by adsorption process. Most of the previous works were accompanied by several flaws in their theoretical and experimental parts. The present work hopefully will help in resolving several shortcomings of the earlier studies.

The preceeding studies on the removal of hexavalent chromium from the wastewater gave some qualitative and quantitative information. However, there are significant deficiencies in them. Some of well established methods have been in practice for decades such as precipitation, co- precipitation, reverse osmosis, ion exchange. The processes as precipitation and co-precipitation simply remove chromium from wastewater by reduction. Although these technique are quite satisfactory in term of purging out chromium and other heavy metals, but they produce solid waste residues containing toxic compounds whose disposal is generally by land filling and which could cause possible ground water contamination. Reverse osmosis and ion exchange techniques are expensive and may not be suitable for all the application.

A second deficiency in the more recent investigations is the activated carbon adsorption systems which are widely used and has played an important role in cleaning up industrial and municipal wastewater. The various researchers (Aggaerwal, *et al.*, 1999; Deepak, *et al.*, 1991; Mahesh, *et al.*, 1999; Ramos, *et al.*, 1994; Singh, *et al.*, 1994; Srivastava, *et al.*, 1996; Tiwari, *et al.*, 1989; Huang, *et al.*, 1977; Alerts, *et al.*, 1989; Kim, *et al.*, 1977; Lee, *et al.*, 1989) investigated use of activated carbon in their studies. The investigators have used various grades of carbon after subjecting them to different activation procedures to prepare it for the studies. The activated carbon is the most common commercial adsorption medium and full-scale processes based on this are in operation in developed countries. However, use of activated carbon is not suitable for developing counties like India because of its high cost and losses during the regeneration restricts its application. Hence the commercial activated carbon could substitute with unconventional, low cost and locally available agricultural waste adsorbents.

Several researchers have explored the possibility of using agricultural waste adsorbents such as sugarcane bagasse, coconut jute, nutshell, rice straw, rice husk, waste tea leaves, groundnut husk, crop wastes, peanut hulls, compost wates etc. for the removal of Cr (VI) by low cost adsorbents (Gang, *et al.*, 1999; Deo, *et al.*, 1992; Huang, *et al.*, 1975; Periasamy, *et al.*, 1991; Ayub, *et al.*, 1998, 1999, 2001, 2002; Drake, *et al.*, 1996; Shukhla and Sakhardane, 1991; Weber, 1996; Chand, *et al.*, 1994; Siddiqui, *et al.*, 1994; Camino, *et al.*, 2000; Orhan, *et al.*, 1993; Okiemen, *et al.*, 1991; Maranone, *et al.*, 1992 ; Arulnatham, *et al.*, 1989 Alaerts, *et al.*, 1989). Unfortunately none of the investigators have thoroughly studied the nature of adsorbents and its characteristics using electronic microscope scanning. In the present studies, the characteristics of the adsorbent before and after the adsorption were studied first time. Few researchers used adsorbents without initial treatments, which ultimately increased BOD (Biochemical Oxygen Demand) due to hydrolysis of lignin. Some agricultural waste adsorbents such as rice straw, rice husk, wheat straw etc. clogged the adsorption column due to large increase in volume after getting soaked in water. Initial characterization of the adsorbents was in complete in the most of studies.

Another shortcoming in the past investigations is that only batch studies were conducted to find the effects of pH, contact time, adsorbents dose etc. Results obtained in the study were not properly interpreted (Chand, *et al.*, 1994; Gang, *et al.*, 1999; Deo, *et al.*, 1992). Kinetic studies have not been performed to understand mechanism of adsorption process. Chand, *et al.* (1994) carried out batch experiments to asses the suitability of bagasse and coconut jute for the removal of Cr (VI). The author conducted batch studies and have not mentioned how the adsorbents were prepared. Further more, desorption/leaching testes were not conducted. Removal of heavy metals using low cost adsorbents was also studies by Rao, *et al.* (2000) and Beiley *et al.* (1999). Wheat straw dust, coconut jute, bark, peat, seaweeds etc. were used as adsorbents by them in the batch studies. Column studies were not conducted to compare the results with batch performance. No initial and final characterization of the adsorbent was reported.

Rai, (1995) has taken bagasse ash, brick kiln ash, fly ash, Rice husk ash and saw dust as adsorbents for their batch studies and found that the adsorption equilibrium data for different adsorbents fitted well to Freundlich and Langmuir isotherms. The rice husk ash and saw dust were found best and most inexpensive adsorbents for chromium removal and he recommended column studies for optimizing the process conditions in his research work. Cimino, *et al.* (2000) have used hazelnut shell as biosorbent substrate for the removal of ions such as Cd^{2+}, Zn^{2+}, three- and hexavalent chromium in batch equilibrium tests. The Cr (VI) removal was found to be pH dependent and fitted with the Langmuir isotherm model. The authors did not make column study as well as leaching test in their study.

As explained above, the removal of hexavalent chromium by adsorption has been investigated by several workers in the past. However, there are several shortcomings and attention has not been paid to make a complete investigation. In the present work, for the first time, efforts were made to fulfill following objective.

1. The various processes available for the chromium (VI) removal have been evaluated and thrust has been laid to develop environment friendly, cost effective appropriate technology.
2. In present study, adsorption technique for the removal of heavy metals has been adopted. It is a tertiary treatment process and is suitable for treating domestic and industrial waste, due to its sludge free clean operation.
3. Due to the high cost of activated carbon and its complex regeneration process, the agricultural waste adsorbents such as neem bark, coconut shell and sugarcane bagasse have been investigated in the study.
4. The adsorbents were characterized before and after the adsorption using scanning electron microscopy technique. The adsorbents were pulverized, screened and stock of uniform size was prepared to use in the study. Thus effect of particle size on the adsorption was studied first time in a systematic fashion
5. Column studies were made to determine the parameters to design a continuous system. Lastly leaching studies on each adsorbent was made to understand the hydrolysis of the agricultural waste material under the experimental conditions. This was done first time and none of the past investigator carried out this important output.

Chapter 5
Materials and Methods

5.1 Collection of Samples and Site Description

Aligarh is a medium sized semi industrialized town located in northern India 135 Km South – East of New Delhi. The city is famous for the electroplating of locks and other building materials for the indigenous as well as export market. A number of water samples from various parts of the town were analyzed to identify the contaminated surface waters. The samples of electroplating effluents were also collected from the discharge points of the factory/surroundings and its nearby and stored as per standard methods before analyzing the physico-chemical parameters. All the samples were tested for pH, chloride, alkalinity, suspended solids, hardness, total dissolved solids and heavy metals chromium contamination. The study reveals that most of the places the surface water is within the limit except a few places where the water was observed to be hard and contained chromium (VI) contamination. The details are given in Table 5.1. However the chromium contamination in the surface water was found above the prescribed safe limit 0.1 mg/l (Table 2.6).

5.2 Selection of Adsorbents and Sample Preparation

In the present work various agro-based waste materials such as, coconut shell, neem bark (azadichta Indica) and raw bagasse have been used as adsorbents for the investigating the removal of Chromium (VI) from the wastewater using batch and column adsorption processes. These adsorbent are widely available in India. A considerable amount of different types of agricultural wastes are generated during the harvesting of various crops. These agricultural wastes if properly utilized can serve as a means of pollution control from various industrial units. Neem bark are widely available all over the India and coconut shell is a waste material easily available in the southern states. Chand, *et al.* (1994) also used raw bagasse for the treatment of hexavalent chromium without the characterization of adsorbent and carried out only effect of adsorbent dose, pH, Initial concentration, temperature and isotherm parameters in study. Although other investigators have studied removal

Table 5.1: Quality Parameters of Surface Water of Aligarh Town

Sl.No.	Place	pH	Chloride mg/l	Alkalinity mg/l as $CaCO_3$	Suspended Solids mg/l	Hardness mg/l as $CaCO_3$	Total Solids mg/l	Chromium (VI) mg/l
1.	Anupshahr road	7.6	77.2	52.6	285	395.5	957.0	46.8
2.	Aligarh Industrial area	6.50	105.58	–	326	640.40	1489	51.5
3.	Ramghat road	7.34	123.5	51	1.23	245.50	174	2.3
4.	Jivangarh interior	6.80	155	–	487	290.5	2123	5.87
5.	Upper fort	6.75	132.2	75.4	574	745.25	1543	54.8
6.	Bus stand	7.0	142.3	53.5	369.5	342.0	2343	6.25
7.	Exhibition ground	7.9	160.5	70.2	415.6	254.32	1985	3.38
8.	Rasalganj	6.7	153.5	80.3	567.7	289.30	2430	39.55

Safe permissible limit of Chromium (VI) is 0.1 mg/l (Federal Register 1981).

of metals such as Cadmium and Copper using sugarcane bagasse, only Chand, *et al.* (1994) have used for the removal of chromium (VI). In this investigation, sugarcane bagasse was studied to verify the results of Chand, *et al.* (1994). Secondly Chand, *et al.* (1994) did not characterize the bagasse before and after adsorption and do not carried out the column study. No other investigator has carried out adsorption on Neem bark and coconut shell for the removal of Cr (VI). As both materials are abundantly available throughout India, so it was considered to study the adsorption characteristics of both of them. Some investigator such as Arulnatham, *et al.* (1989) has studied removal of cadmium and lead on coconut shell but none of them have used Neem bark.

5.2.1 Coconut Shell

The coconut shell was first dried at a temperature of 150° C for 5 hours. After grinding it was sieved to obtain average particle size of 225 mesh (Indian Standard Sieve). It was then washed several times with distilled water to remove lighter materials and other impurities. The adsorbent was dipped in 1N NaOH for a period of 10 hrs and washed several times with distilled water to remove the lignin content and then dried. The adsorbent was again washed separately with double distilled water two to three times and dipped into 0.1N H_2SO_4 for the period of 10 hrs to remove traces of alkalinity. The acid treated adsorbent is washed thoroughly with double distilled water. Thereafter the material was dried in sun and stored in a desiccator.

5.2.2 Neem Bark

Fresh neem bark (*Azadichta indica*) was obtained from neem tree. Then it was dried at a temperature of 100° C for 5 hours. After grinding it was sieved to obtain average particle size of 225 mesh (Indian Standard Sieve) to increase the surface area. It is then washed several times with distilled water till it was free from colour causing substances and supernatant was clear. The adsorbent was dipped in 1N NaOH for a period of 10 hrs and washed several times with distilled water to remove the lignin content and then dried. The adsorbent was again washed separately with double distilled water two to three times and dipped into 0.1N H_2SO_4 for the period of 10 hrs to remove traces of alkalinity. The acid treated adsorbent was later washed thoroughly with double distilled water. Thereafter the material is dried in sun and stored in desiccator.

5.2.3 Sugarcane Raw Bagasse

In the present study the inner part of the sugarcane bagasse is taken as adsorbent. Washing was done by putting the bagasse in beaker, shaking it several times till it removed the dust and soluble impurities. After crushing and grinding it was sieved to obtain average particles of 225 mesh (Indian Standard Sieve). The adsorbent was dipped in 1N NaOH for a period of 10 hrs and washed several times with distilled water to remove the lignin content and then dried. The adsorbent was again washed separately with double distilled water two to three times and dipped into 0.1N H_2SO_4 for the period of 10 hrs to remove traces of alkalinity. The acid treated adsorbent was washed thoroughly with double distilled water till the wash water was colourless. Thereafter the material was dried in sun and stored in a desiccator.

5.3 Chemicals

Chromium (VI) was used as an adsorbate. In order to have waste of uniform characteristics and to avoid interference with other impurities the laboratory wastewater was prepared by dissolving a known amount of potassium dichromate in a known volume of distilled water. For 1000 mg/l hexavalent-chromium concentration solution 2.282g $K_2Cr_2O_7$ (AR Grade) was dissolved in 1.0 L of distilled water. The chemicals used in the present study were: Potassium dichromate (Qualigens Fine Chemicals, Mumbai, India), Sodium hydroxide and Sulphuric acid (Qualigens Fine Chemicals, Mumbai, India) all laboratory grade. The stock solution was stored in a safe location and the required quantity was diluted with the deionized double distilled water to make solution of concentration used in the studies. The water used in the study was deionized and later double distilled and stored securely to avoid any contamination.

Electroplating rinse wastewater was obtained from Sigma Engineering works, Anoop shaher road, Aligarh. The adsorbate concentrations were in the range of the Cr (VI) concentration found around Aligarh city as mentioned in Table 2.5.

5.4 Experimental Procedure

In order to understand the adsorption behavior a number of batch studies were conducted according to the Standard Method to investigate the effect of adsorbent dose and contact time, pH, concentration of metal, particle sizes and temperature variation. For these studies, wastewater of various concentrations of Cr (VI) was prepared from the stock solution and kept separately in glass stoppered conical flasks. Then suitable doses of adsorbent were added to the wastewater in a conical flask of 250 ml. The system was equilibrated by shaking the contents of the flasks at room temperature on a mechanical shaker (Indian Scientific Instruments factory, Ambala Cant, India) so that adequate time of contact between adsorbent and the metal ion was maintained. The suspension was filtered through Whatman (No. 1) filter paper and the filtrate was analyzed to evaluate the concentration of Cr (VI) metal in the treated wastewater by using atomic absorption spectrophotometer (GBC 902). Adsorption studies were made for various times. Ultimate saturation time was also determined for each dose. These studies were carried out at the room temperature which varied from 20-25° C in winter time to 30-38° C in summer time. In the elevated temperature studies the temperature was maintained by external heaters in closed chamber.

5.4.1 Batch Studies

In order to evaluate the effect of adsorbent dose and contact time, pH, concentration of metal, particle sizes and temperature variation on the potential of agricultural waste adsorbents to remove Cr (VI) from the wastewater, batch experiments were designed as described below.

5.4.1.1 Effect of Adsorbent dose and Contact time on Fraction Adsorbed

To study the effect of various adsorbent dose and contact time on Cr (VI) removal, suitable doses of adsorbent ranging from 2.0 to 60 g/l were added to 50 ml wastewater sample having initial Cr (VI) concentration of 50 mg/l. The initial pH of the wastewater before addition of the adsorbent was recorded as 1.5 for coconut shell and sugarcane bagasse, where as 3.0 for neem bark. The final concentration of Cr (VI) in the solution was determined after (0.5 hrs to 24 hrs) contact time. Thus the amount of Cr (VI) adsorbed at various adsorbent doses and contact time has been determined as fraction of chromium adsorbed at any time divided by the chromium adsorbed at the saturation time.

5.4.1.2 Effect of pH on Fraction Adsorbed

The pH of a solution from which adsorption occurs may affect the extent of adsorption for several reasons. Due to strong adsorption of hydrogen from the solution and hydroxyl ions on the adsorbents, the adsorption of other ions is strongly influenced. The pH of the solution also affects the degree of ionisation, and in turn affects the extent of adsorption. In the study various pH values ranging form 1.0 to 9.5 were studied at an initial Cr (VI) concentration of 50 mg/l. An adsorbent dose of 10.0 g/l was added to 50 ml of wastewater. The pH was maintained by 1M sulphuric acid/sodium hydroxide. The amount of Cr (VI) adsorbed was determined as described in the above section.

5.4.1.3 Effect of Initial Cr (VI) Concentration on Fraction Adsorbed

Wastewater samples having different concentration of Cr (VI) ranging 5.0 to 100 mg/l were taken separately in glass stoppered conical flasks, and 10 g/l adsorbent was added to each flask. The size of wastewater sample was taken as 50 ml in each case and the contact time was kept 2.5 hrs for sugarcane bagasse, 24 hrs for coconut shell and 12 hrs for neem bark. The temperature was maintained as 30° ± 1° C.

5.4.1.4 Effect of Contact Time and Particles Size on Fraction Adsorbed

Uptake of adsorbate species is rapid in the initial stages of the contact period and becomes slow near equilibrium. Three particles size 75 micron, 150 micron and 300-micron sieve (Indian standard Sieves) with an initial adsorbate concentration of 50mg/l and the contact time from 0.5 to 24 hrs were studied. The initial pH of the wastewater before addition of the adsorbent was measured as 1.5 for coconut shell and sugarcane bagasse, where as 3.0 for neem bark. The temperature was maintained at 30.0±1°C in all the runs except those meant for investigating the effect of temperature.

5.4.1.5 Effect of Contact Time and Temperature on Fraction Adsorbed

Adsorbate concentration was maintained at 50 mg/l and adsorbent dose of 10.0 g/l in all experiments. The effect of temperature upon the adsorption rate was investigated at 30°C, 40°C and 50°C by keeping them in temperature controlled oven (Indian Scientific Instruments factory, Ambala Cant, India). The contact time varied from 30 minutes to 24 hrs and the pH was maintained by sulphuric acid and sodium hydroxide. The residual concentration of chromium was subsequently determined.

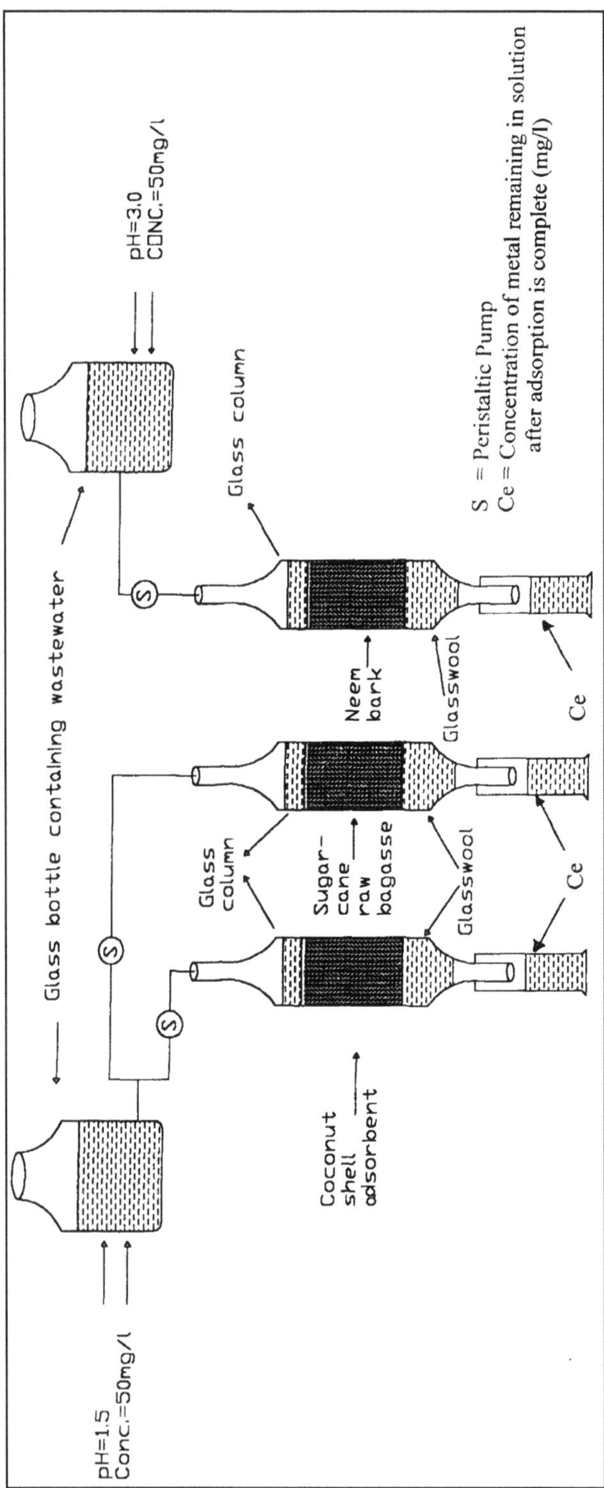

Figure 5.1: Experimental Setup for Column Study.

5.4.2 Column Studies

Although batch laboratory adsorption studies provide useful information on the application of adsorption to the removal of waste constituents, column study provide the practical application in the waste treatment technology for the design of continuous adsorption columns. When wastewater is introduced at the top of clean bed of adsorbent, most solute removal initially occurs in a rather narrow band at the top of the column, referred to as the adsorption zone. As adsorption continues, the upper layers of packing material become saturated with solute and adsorption zone progress downward through the bed. Eventually, the adsorption zone reaches the bottom of the column, and the solute concentration in the effluent begins to increase. A plot effluent solute concentration verses time usually yields an S- shaped curve, referred as a breakthrough curve. The point on the S-shaped curve at which the solute concentration reaches its maximum allowable value is referred to as breakthrough.

Column studies were conducted using a glass column (Internal diameter 1.0-cm). Adsorbent was suspended in distilled water for 15 minutes by shaking in a beaker at the speed of 150 rpm, and then transferred in to the glass column. This was done to disperse the particles properly and to avoid conglomeration. After this, particles were transferred in the column. The glass wool was kept at the bottom and top ends in order to avoid its loss with the liquid flow or floating. The flow rate was maintained at 1.0 l/d. Concentration in the influent and exit stream from the column were determined using atomic absorption spectrophotometer (GBC 902). The experimental set-up is shown in Figure 5.1. The column experiments were conducted at the room temperature.

5.4.3 Hydrolysis of Adsorbent and Desorption

It is likely that the agricultural waste materials hydrolyze at the experimental pH. Therefore it was found necessary to study the hydrolysis characteristics of the adsorbents. Desorption tests were conducted for all the adsorbents after their use in the equilibrium adsorption studies. 10 gram of saturated adsorbent was placed in 250 ml stoppered conical flask with distilled water and was kept at room temperature for over 3 hours. After this the adsorbent was filtered and the solution was analyzed for the chromium content.

Chapter 6

Results and Discussions

6.1 Coconut Shell

6.1.1 Batch Study Parameters

The experimental data obtained during the adsorption studies are being used to evaluate the effect of adsorbent dose and contact time, influence of pH, effects of various initial Cr (VI) concentrations, effect of variation of contact time and different particle sizes, effect of contact time and temperature. Adsorption isotherms are plotted to determine the feasibility of adsorbing system. Kinetic studies have been performed to understand the mechanism of adsorption. Thermodynamics parameters indicate the feasibility of the process. Column studies have been carried out to compare these results with the batch studies.

(a) Effects of Adsorbent Dose and Contact Time on Adsorption

The response of adsorbent dose and contact time on the removal of Cr (VI) is shown in Figure 6.1. The chromium uptake was studied by varying the amount adsorbents from 5 to 60 g/l and contact time from 1 hour to 24 hours. The pH was maintained at the value of 1.5 which it was found that maximum Cr (VI) removed was taking place. The results show that an increase in the fraction adsorbed of Cr (VI) occurs with corresponding increase in the dose of coconut shell and contact time up to certain level, beyond which the fraction adsorbed remains constant. It is evident that a dose of 10g/l is sufficient to remove 55 – 75 per cent Cr (VI) in 1-12 hours. The increase in the removal efficiency with simultaneous increase in adsorbent dose and contact time is due to the increase in surface area and hence more active sites are available for the adsorption of Cr (VI). Similar results were reported by (Bansal and Sharma, 1992; Kim and Joltech, 1977; Singh, *et al.*, 1992; Ayub, *et al.*, 2001,2002) for the removal of chromium compounds by adsorption.

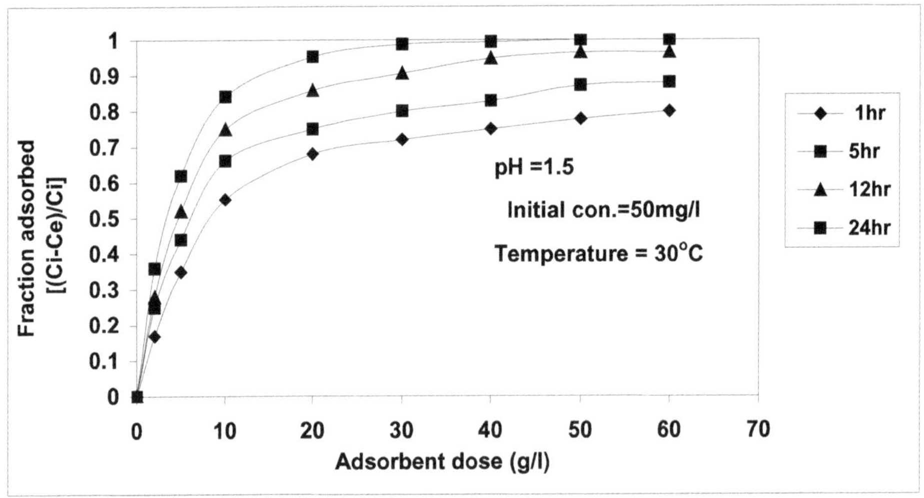

Figure 6.1: Effect of Adsorbent Dose and Contact Time on Adsorption.

(b) Effect of pH on Adsorption

Influence of pH on fraction adsorbed was studied by varying the pH of the test solution by adding H_2SO_4 or NaOH solution of known strength is shows in Figure 6.2 The pH of the wastewater primarily affects the degree of ionization of the adsorbate and surface properties of adsorbent. This subsequently leads to a shift in reaction kinetics and equilibrium characteristics of adsorption process. The adsorption of various anionic and cationic species on such adsorbents has been explained on the basis of the competitive adsorption of H^+ or OH^- ions along with adsorbate molecules. In case of adsorbents having metallic oxides, it is the common observation that the surface adsorbs anions favorably in low pH range

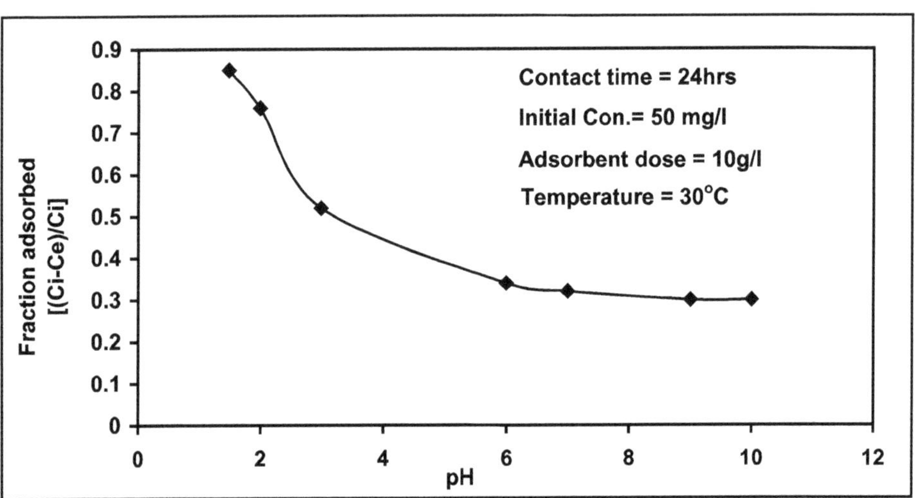

Figure 6.2: Effect of pH on Fraction Adsorbed.

due the presence of H^+ ions (Gebehard and Coleman, 1974) whereas, the surface is active for the adsorption of cations at higher pH values due to the accumulation of OH^- ions (Huang and Stunm, 1973). The results show the maximum removal efficiency (85 per cent) is observed at pH 1.5 while on increasing the pH value it decreases. About 32 per cent removal efficiency is recorded at pH 7. Sharma, *et al.* (1993) reported that sorption of chromium was highly pH dependent and the best results were obtained in the pH range 1.5 – 3.0. Huang, *et al.* (1975) found 100 per cent chromium removal at pH 2.0 when the concentration was less than 5.2 mg/l as chromium.

(c) Effect of Initial Cr (VI) Concentration on Adsorption

The effect of initial Cr (VI) concentration on fraction adsorbed (Figure 6.3) were studied over the wide range of chromium concentration (5-100 mg/l). It may be observed that the chromium uptake is rapid during the initial period of adsorption and the maximum removal (100 - 90 per cent) is achieved at 5-20 mg/l concentration. The removal efficiency of chromium decreases when chromium concentration is increased. However the removal efficiency is recorded as 84.3 per cent at a concentration of 50 mg/l. This is in agreement with the observations made by earlier workers (Bansal and Sharma, 1982; Haribabu, *et al.*, 1993a,b; Grover, *et al.*, 1982; Gupta, *et al.*, 1992; Khanna and Malhotra 1977; Kumar, 1987; Kumar, *et al.*, 1987; Mall, 1992; McKay, *et al.*, 1991; Singh and Mishra, 1990; Singh, *et al.*, 1994; Chand, *et al.*, 1994). The adsorbate concentration and contact time between the adsorbate and adsorbent are of great significant in the treatment of wastewaters by adsorption. A rapid uptake of pollutants and establishment of equilibrium in a short period of time signifies the efficiency of the adsorbent. The initial solute concentration also plays a vital role in making a first guess of the required dose, as a given mass of the adsorbent can adsorb only a fixed amount of solute. Therefore, the more concentrated the solution, the smaller is the volume of solution or effluent that a mass of adsorbent can purify. The fractional adsorption is low in the high concentration range (McKay, 1982). The time required to reach equilibrium is longer for more porous adsorbents like activated carbon and polymeric materials and shorter in the case of relatively less porous or non-porous adsorbents such as brick kiln ash, china clay, fly ash, china clay, fly ash and wollastonite, etc. This happens as the porous materials have higher surface area compared to the hard surfaces and the adsorbate ions have to diffuse through the pores before they could come in contact with the active sites.

In the physical adsorption most of the adsorbate species are adsorbed on the solid-solution interface within a short period of time (Stumm and Bilinski, 1972). However, strong chemical bonding of adsorbate with adsorbent requires a longer contact time for the attainment of equilibrium. A review of the available literature on adsorption at the solid-solution interface reveals that the uptake of adsorbate species is fast during the initial stage of contact and becomes slower near the equilibrium. This is obvious from the fact that a large number of surface active sites are available for adsorption during the initial stages and after a lapse of time, the remaining surface sites are difficult to be occupied because of the repulsion between the solute molecules at the solid surface and those in the bulk of the liquid. It is also likely that the mass transfer through the pores become the controlling factor in the process.

Initial concentration is found to have no significant effect on the equilibrium time for the adsorbents used. Similar observations were reported by earlier workers also for the adsorption of chromium, dyes, phenols etc. on activated carbon, china clay, fly ash, rice husk ash, saw dust, brick kiln ash, paddy straw, wollastonite, etc. (Gupta, *et al.*, 1992; Kumar, 1987; Kumar, *et al.*, 1987; Pandey, *et al.*, 1985; Sharma, *et al.*, 1991).

In an aqueous system, the water molecules surrounding these oxides play an important role in the adsorption of chromium. The electrical charges and potential due to the electro-chemical properties of the two phases at the solid-solution interface become equal. This results in the formation of an electrical double layer which is greatly influenced by the physico-chemical properties of the bulk phase (Mall, 1992; Waldsax and Jaycock, 1971). During the adsorption process, aqua-complexes of the metal ions are formed which are amphoteric in character and may either exchange protons from water molecules or may donate protons to hydroxyl ions in following manner.

$$M^{n+}(OH)_n(OH_2)_m \rightarrow \left[M^{n+}(OH)_{n+1}(OH_2)_{m-1}\right]^{-1} + H^+ \qquad 6.1$$

$$M^{n+}(OH)_n(OH_2)_m + H^+ \rightarrow \left[M^{n+}(OH)_{n-1}(OH_2)_{m+1}\right] \qquad 6.2$$

Co-ordination number of the metal ions is assumed to remain constant in the above equilibria. During the process of contacting aqueous solutions of chromium with the oxides of metals and non-metals present in adsorbents, the formation of oxide-water interface takes place and H^+ and OH^- ions become the constituent parts of this interface and there exists a state of thermodynamic reversibility (with respect to these ions) at the surface of adsorbent. The H^+ and OH^- ions are then the potential determining ions for mineral oxides (Fuerstenu, 1970) with respect to the resulting hydroxylated oxide surfaces in contact with aqueous solutions. The progress of adsorption is governed by the kinetics of surface hydroxylation and subsequent acid-base dissociation of aqua-complexes of the oxides, yielding there by, an oxide-solution interface in the following manner (Ahmed, 1972; Gupta, *et al.*, 1990; Mall, 1992; Panday, 1984).

$$M^{\mapsto O}_{\to OH} \quad \underleftrightarrow{OH^-} \quad M^{\mapsto OH}_{\to OH} \quad \underleftrightarrow{H^+} \quad M^{+}_{\to OH} \qquad 6.3$$

It can be seen from the above equilibria that the negatively charged ion density lies on the oxygen site of the surface, where as the positively charged ion density lies on the metal site of the oxide surface (Ahmed, 1972; Gupta, *et al.*, 1990; Mall, 1992; Panday, 1984; Levine and Smith, 1971) also suggested a similar mechanism using a double layer model for adsorption of anions at the oxide-solution interface. The above mechanism explains satisfactorily the variation in adsorption with contact time between adsorbent and adsorbate on the basis of the surface reaction of adsorbate species with interface, which ultimately reaches the equilibrium condition.

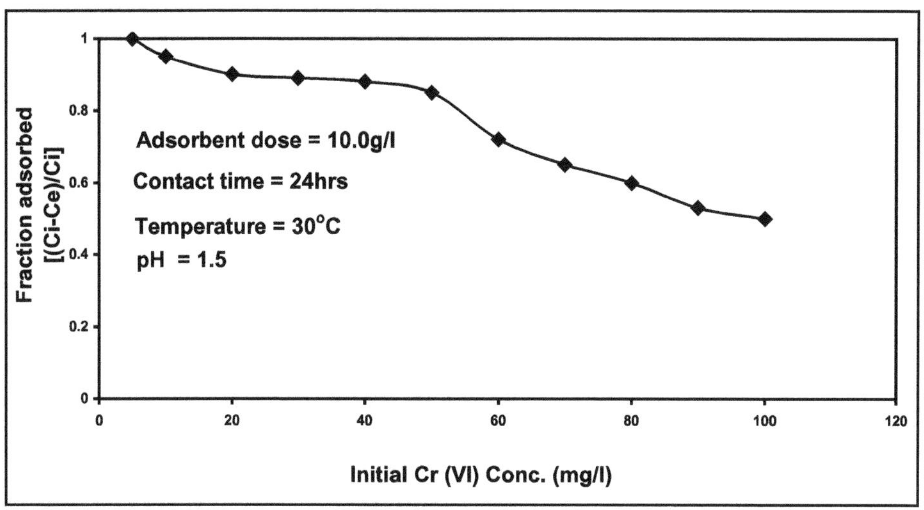

Figure 6.3: Effect of Cr (VI) Concentrations on Fraction Adsorption.

Wright and Hunter, (1973) used a double layer approach and proposed another mechanism for explaining the adsorption of ions at the hydroxide-solution interface. According to them the following equilibria exist at the surface and in the bulk of the solution which seem to be more appropriate in explaining the surface reaction involved in such oxide-solution systems.

$$MOH_s \Leftrightarrow MOH_b \qquad 6.4$$

$$MOH_b + H^+ \Leftrightarrow M_b + H_2O \quad \text{(in the bulk)} \qquad 6.5$$

$$MOHs + H_b^+ \Leftrightarrow M_s^+ + H_2O_b \quad \text{(on the surface)} \qquad 6.6$$

Where MOH stands for any multivalent metal hydroxide.

(d) Effect of Contact Time and Particle Size on Adsorption

The adsorbent particle size has significant influence on the adsorption time and kinetics of adsorption. The effect of particle size gives important information on achieving optimum utilization of adsorbent and on the nature of breakthrough curves for designing packed bed adsorbers. Three particle sizes 75 micron, 150 micron and 300 micron sieve (Indian standard Sieves) under optimal conditions of adsorbent dose, pH and contact time with an initial adsorbate concentration of 50 mg/l were studied. Figure 6.4 show fraction adsorbed versus contact time indicating that the fraction removed decreases with an increase in adsorbent particle size. This may be explained on the basis of the active surface area available for the adsorption, which is greater for small particle sizes (Poots and Healy, 1976). Finer adsorption material offers a significantly lesser mass transfer diffusion resistance in the micro pores.

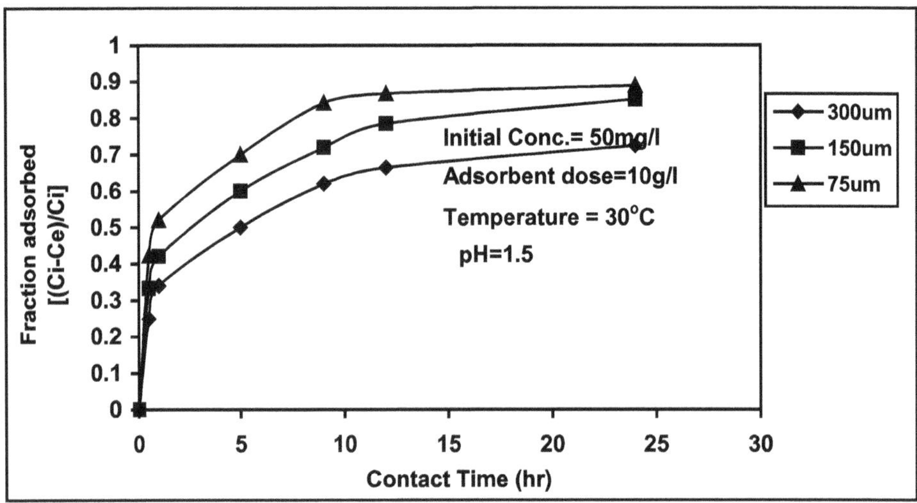

Figure 6.4: Effect of Contact Time and Particle Size on Adsorption.

(e) Effect of Contact Time and Temperature on Adsorption

The effect of temperature upon the adsorption rate was investigated at 30°C, 40°C and 50°C. The results are shown in Figure 6.5. It is observed that the mass of the chromium adsorbed per unit mass of adsorbent increases with increasing temperature. This shows that the adsorption process is endothermic in nature. Several earlier investigators (Khare, *et al.*, 1987; Knocke and Hemphill, 1981; Panday, 1984; Panday, *et al.*, 1985, 1986; Singh, *et al.*, 1988) have reported increase in uptake of adsorbate with increase in temperature. The effect of temperature on the rate of adsorption is used for determining of the energy of activation. The dependency of

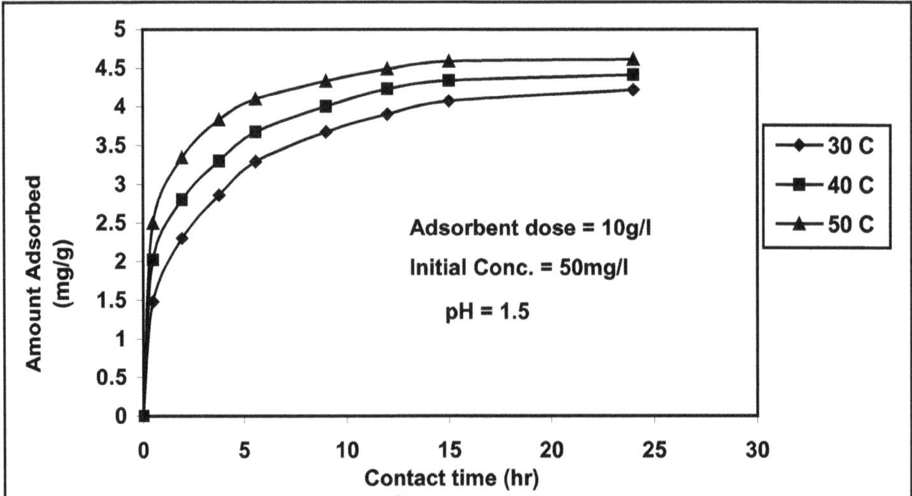

Figure 6.5: Effect of Contact Time and Temperature on Adsorption.

specific rate constant K_{ad} on temperature for the over all removal process follows the Arrhenious equation as given below.

$$K_{ad} = A\,e^{-\left(\frac{E}{RT}\right)} \qquad 6.7$$

$$\ln K_{ad} = \ln A - \frac{E}{RT} \qquad 6.8$$

where, A, is the frequency factor. It is nearly independent of temperature and represents for the available adsorption sites. The value of A and E can be calculated from the plot of ln K_{ad} versus 1/T. This will be discussed for the present systems in the section (h) Kinetics of Adsorption. (Rai, 1995).

(f) Adsorption Isotherms

Experimental evaluation of isotherm constants is done to determine the feasibility of an adsorbing system. Batch experiments were performed to obtained data. Such data were obtained for both laboratory and actual wastewater and were analyzed using the Freundlich and Langmuir models (De Castro *et al.*, 2001; Ahmad and Ram, 1992; Haribabu, 1993; Mall, 1992; Sharma, *et al.*, 1990).

Langmuir isotherm can be represented by the following mathematical relationship.

$$\frac{x}{m} = \frac{abCe}{1 + aCe} \qquad 6.9$$

$$\frac{1}{x/m} = \frac{1}{b} + \frac{1}{abCe} \qquad 6.10$$

where,

x = Amount of material adsorbed (mg)

m = Weight of the adsorbent (mg)

C = Concentration of material remaining in solution after adsorption is complete (mg/l)

a and b = Empirical constants

a linear plot is obtained when the quantity 1/(x/m) is plotted against 1/C. The constants a and b can be determined from the slope and intercept of the plot (Figure 6.6). The essential features of the Langmuir isotherm can be expressed in terms of dimensionless separation factor R_l.

$$R_l = \frac{1}{(1 + bCi)} \qquad 6.11$$

where Ci is the initial concentration and b is the Langmuir constant, separation factor R_l indicates the isotherm shape and its values are given in Table 6.1.

If $R_l > 1$ the nature of adsorption is unfavorable

$R_l = 1$ the nature of adsorption is linear

$0 < R_l > 1$ the nature of adsorption is favorable

$R_l = 0$ the nature of adsorption is irreversible

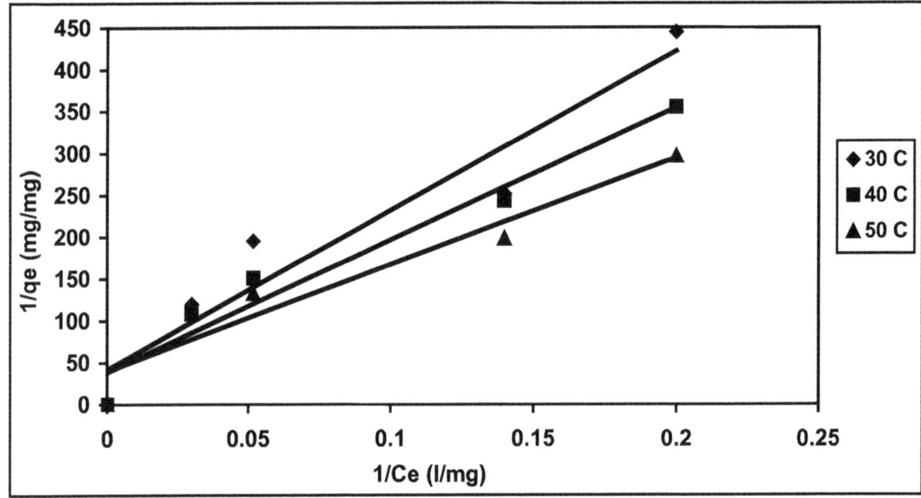

Figure 6.6: Langmuir Adsorption Isotherms Plot.

Freundlich adsorption isotherm model is expressed as,

$$qe = \left(\frac{x}{m}\right) = K(Ce)^{1/n} \qquad 6.12$$

The above equation is also referred as Van - Bemmelen equation. Fitting into the logarithmic form (Mahesh, *et al.*, 1999)

$$\log\left(\frac{x}{m}\right) = \log K + \frac{1}{n}\log Ce \qquad 6.13$$

$$\log\left[\frac{(Ci - Ce)}{Ci}\right] = \log K + \frac{1}{n}\log Ce \qquad 6.14$$

Table 6.1: Values of Langmuir and Freundlich Isotherms Constants

Temperature	*Langmuir Constants*				*Freundlich Constants*			*Recommended Isotherms*
	a	*b*	*Cc*	R_l	*K*	*1/n*	*Cc*	
30ºC	1900.9	0.0220	0.917	0.476	0.0010	0.555	0.920	$0.0010Ce^{0.502}$
40ºC	1580.6	0.0244	0.954	0.450	0.0011	0.591	0.962	$0.0011Ce^{0.599}$
50ºC	1273.0	0.0315	0.924	0.388	0.0011	0.638	0.954	$0.0011Ce^{0.689}$

where, x/m is the amount of Cr (VI) adsorbed per unit mass of adsorbent (mg/mg) and Ce is the equilibrium concentration of aqueous solution. K is a constant, which is measure of adsorption capacity, and 1/n is a measure of adsorption intensity (Figure 6.7). Langmuir and Freundlich adsorption isotherm constants,

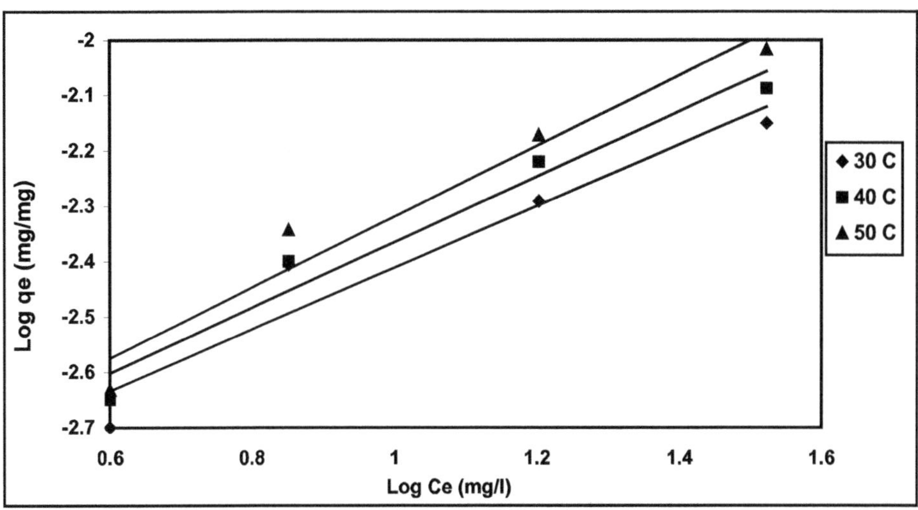

Figure 6.7: Freundlich Adsorption Isotherms Plot.

Coefficient of correlation Cc, Separation factor R_l and recommended isotherms are shown in Table 6.1.

(g) Thermodynamic Parameters

The adsorption at a interface between two phases can be regarded as an equilibrium process. The point of equilibrium can be defined in terms of thermodynamic parameters like free energy, enthalpy and entropy changes. Several researchers (Gupta, *et al.*, 1992; Panday, *et al.*, 1985; Sharma, *et al.*, 1991) have determined these parameters. The positive entropy change for adsorption as observed in some cases may be attributed to an increase in the translational entropy caused by randomness of the displaced water molecules from the surface of the adsorbent (Wright and Pratt, 1974). The negative values of free energy change for a system indicates spontaneity of adsorption process. Adsorption at a solid solution interface generally shows an increase in entropy. This indicates, a faster interaction during the forward adsorption. Association, fixation or immobilizations of adsorbate on the interface between two phases result in loss of the degree of freedom there by, showing a negative entropy effect.

The apparent heat change (ΔH) is related to Langmuir constant b, and should follow the Van't Hoff equation

$$\ln b = \ln b_0 - \frac{\Delta H}{RT} \qquad 6.15$$

Where b_0 is a constant. If the ΔH is positive b should increase with increasing temperature and if the adsorption is exothermic (ΔH negative) b should decrease with increasing temperature

The enthalpy changes of sorption as calculated from the slope of ln b Vs 1/T (Figure 6.8) is found 14.23 Kj/mole. The positive ΔH values confirm the endothermic

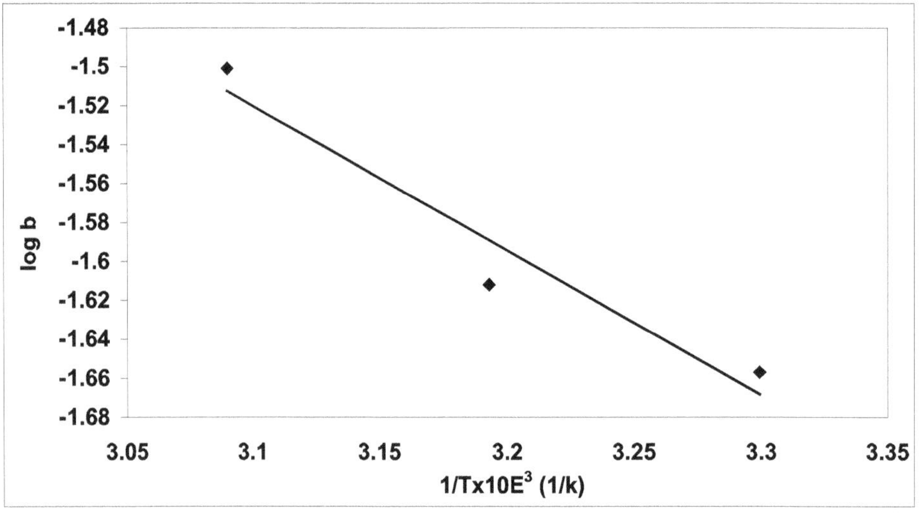

Figure 6.8: Van't Hoff Plot.

nature of the sorption process and suggest the possibility of strong binding between sorbate and sorbent.

The change in free energy (ΔG) and (ΔS) are calculated using the following relationships.

$$\Delta G = -RT \ln\left(\frac{1}{b}\right) \tag{6.16}$$

$$\Delta S = \frac{\Delta H - \Delta G}{T} \tag{6.17}$$

Where T is the temperature in Kelvin and R = 8.31 j/k.mol. The negative value of ΔG indicates the process to be feasible and spontaneous (Singh, *et al.*, 1982; Haribabu, *et al.*, 1993) and positive values of entropy reflected the affinity of the adsorbent material. The values of ΔG and ΔS at different temperature are given in Table 6.2.

Table 6.2: Thermodynamic Parameters at Different Temperatures

Temperature (ºK)	*– ΔG, Kj/mole*	*ΔS, j/mole*
303	9.614	78.693
313	9.662	76.332
323	9.670	74.00

(h) Kinetics of Adsorption

The kinetic modeling for the removal of chromium (VI) by coconut shell has been studied out using the following first order rate expression *i.e.* Lagergren equation

$$\log(qe - q) = \log(qe) - \left(\frac{K_{ad}}{2.303}\right) t \tag{6.18}$$

Where qe and q are the amount adsorbed at equilibrium and any time t respectively. The straight line plot of log (qe-q) versus t for the adsorption shows the validity of lagergren equation and suggest the first order kinetics (Figure 6.9). (Gupta, *et al.*, 2001; Khare, *et al.*, 1987; Knocke and Hemphill., 1981; Mckay, *et al.*, 1980; Haribabu, *et al.*, 1993).

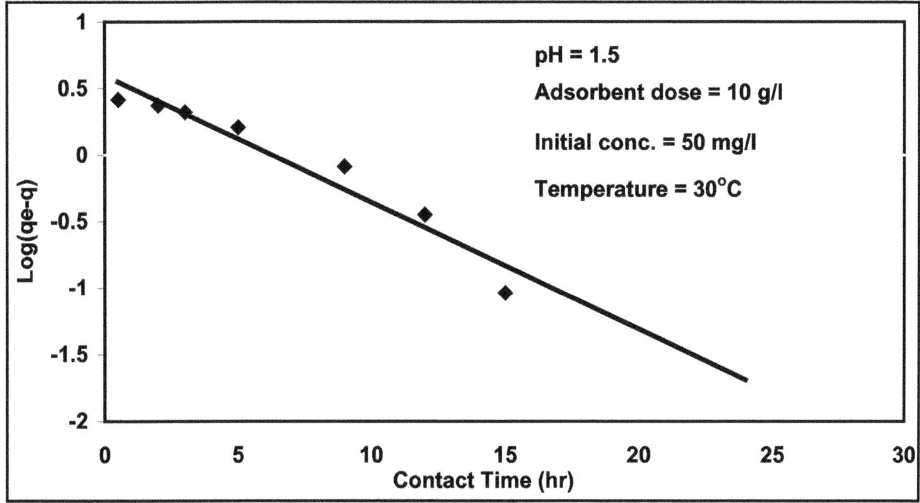

Figure 6.9: Lagergren Plot.

Kinetics of adsorption is the important physico–chemical aspect for the evaluation of parameters describing the adsorption process. It signifies adsorbate uptake rate, and is one of the important characteristics defining efficiency of adsorption. The rate of adsorption also governs the residence time in the adsorption vessel. The rate constant K_{ad} was evaluated using the above equation and is found 0.2194. The value of E as evaluated from the calculated value of K_{ad} using equation 6.8 and was found 1.55 kj/mole.

(i) Interparticle Diffusion Rate

Plot of amount adsorbed per unit mass of adsorbent versus square root of time is shown in Figure 6.10. For unsteady state adsorption a Ficks equation could be used to model on adsorption process and the quantity adsorbed on the solid surface will be proportional to the square root of the time (Bird, *et al.*, 1960). It is evident from this plot that there are two distinct linear sections, the initial steep linear portion and the final relatively flat linear part. The initial linear part indicates that the interparticle diffusion, the latter, less steeper linear, part suggested that adsorption is being controlled by the micro pores. The transport of adsorbate from solution phase to the surface of adsorbent particles is controlled either by film (or surface) diffusion, pore diffusion, pore surface diffusion and adsorption on pore wall or by the combined effect of more than one of these factors (Crank, 1956; Keinath, 1977; Weber, 1972). The film or surface diffusion is an important rate controlling step and is a function of particle size, hydrodynamic conditions, system physical properties, etc. in a well mixed adsorber. The adsorbate species are transported from the bulk to the interior sites of the adsorbent particles and the interparticle transport (within macro and micro-pores) is often the rate limiting step (Huang and Blankenship, 1984; Poots, *et al.*, 1978; Webber and Morris, 1963 a, b, 1964). A functional relationship common to most treatments of interparticle transport is that uptake varies almost linearly with the square root of time, rather than time it self.

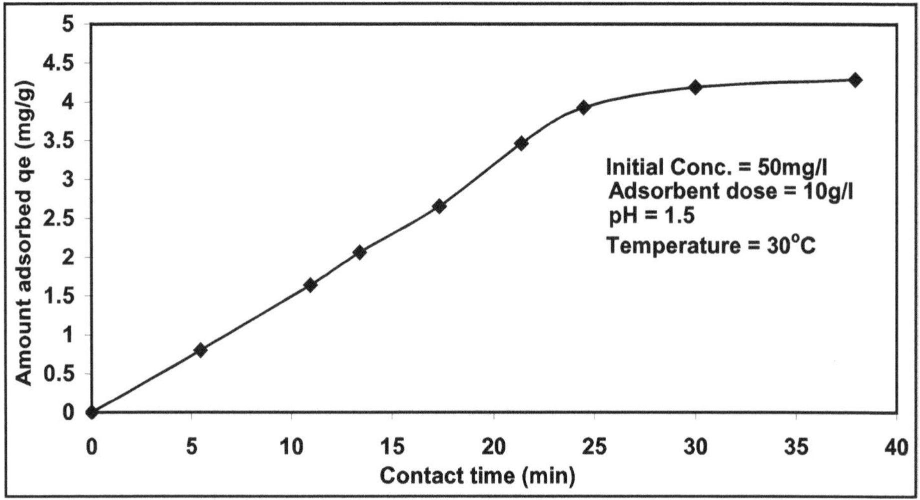

Figure 6.10: Interparticle Diffusion Plot.

6.1.2 Column Study

The breakthrough curve plotted in Figure 6.11. A flow rate of 1.0 l/d having metal concentration of 50mg/l (pH-1.5) was maintained. The column was run till the coconut shell in the adsorption column got exhausted and the treated effluent was analyzed at different time intervals. Column capacities were found greater than the batch capacities due to continuously large concentration at the interface of the sorption zone as the sorbate solution passes through the column while the concentration gradient decreases with time in a batch process (Mall and Upadhyay 1995). Saravanane, *et al.* (2002) used rice husk/saw dust in continuous flow studies. A flow rate of 15 ml/mint. with constant inlet head of 40 cm was maintained, equilibrium was reached within 60 –180 minutes. The breakthrough behavior of Cr (VI), Fe (III) and Hg (II) was studied by Singh, *et al.* (2001) by passing 400 mg/l solution of each metal ion (pH ~ 6) through a glass column (internal diameter 1.25 cm) loaded with 2.0 g active carbon. The flow rate was maintained ~ 2 ml/min.

6.1.3 Desorption and Hydrolysis Tests

Desorption test was conducted for coconut shell after their use in the equilibrium adsorption studies. About 10 g of saturated adsorbent was placed in a 300 ml capacity stoppered BOD bottle with double distilled water and was shaken over a shaker at room temperature for over two hours. After this the adsorbent was filtered and the suspension was analyzed for the chromium content. No chromium was detected in the water. Thus it indicates that adsorbed chromium was not being desorbed.

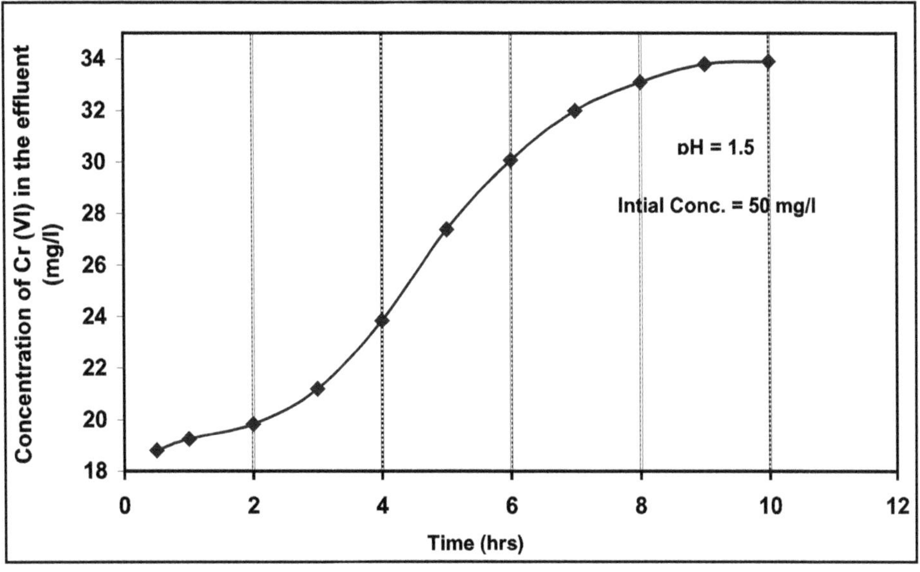

Figure 6.11: Breakthrough Curve.

6.2 Neem Bark (*Azadirichta indica*)

6.2.1 Batch Study Parameters

(a) Effect of Adsorbent Dose and Contact Time on Adsorption

The response of adsorbent dose and contact time on the removal of Cr (VI) is shown in Figure 6.12. Increase in the adsorption occurs with corresponding increase in the dose of Neem bark and contact time up to certain point, beyond which the fraction adsorbed remains constant. It is evident that a dose of 10g/l is sufficient to remove 53.3 – 80.0 per cent Cr (VI) in 1hours-12 hours. The increase in the removal efficiency with simultaneous increase in adsorbent dose and contact time is due to the increase in surface area and hence more active sites are available for the adsorption of Cr (VI).

(b) Effect of pH on Adsorption

Effect of pH on adsorption is shown in Figure 6.13. The results show the maximum removal is observed at pH 3 and on increasing the pH value it decreases. About 75 per cent removal efficiency is recorded at neutral pH 7. The reason for the better adsorption capacity observed at low pH values may be attributed to the large number of H^+ ions present at these pH values, which in term neutralize the negatively charged hydroxyl group (-OH) on adsorbed surface thereby reducing hindrance to the diffusion of dichromate ions. At higher pH, the reduction in adsorption may be possible due to abundance of OH^- ions causing increased hindrance to diffusion of

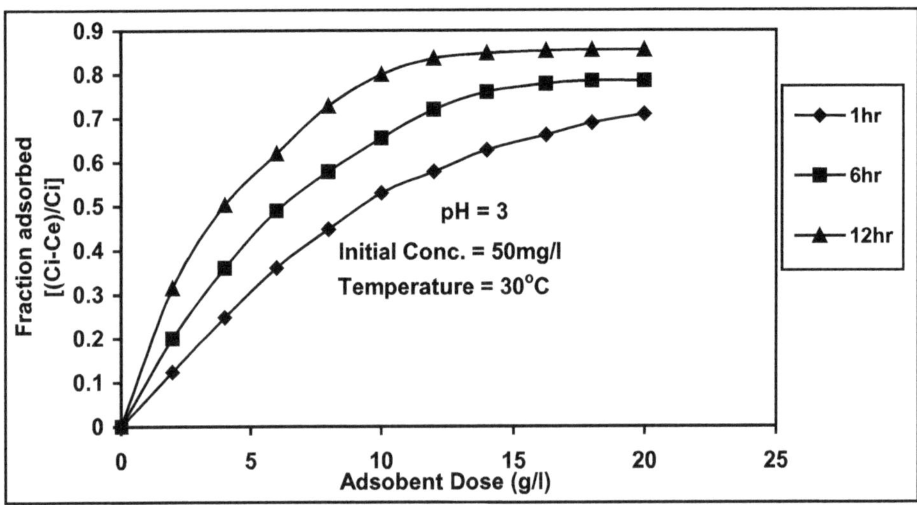

Figure 6.12: Effect of Adsorbent Dose and Contact Time on Adsorption.

positively charged dichromate ions. It is the common observation that the surface adsorbs anions favorably in low pH range due the presence of H^+ ions (Gebehard and Coleman, 1974) where as, the surface is active for the adsorption of cations at higher pH values due to the accumulation of OH^- ions (Huang and stunm, 1973).

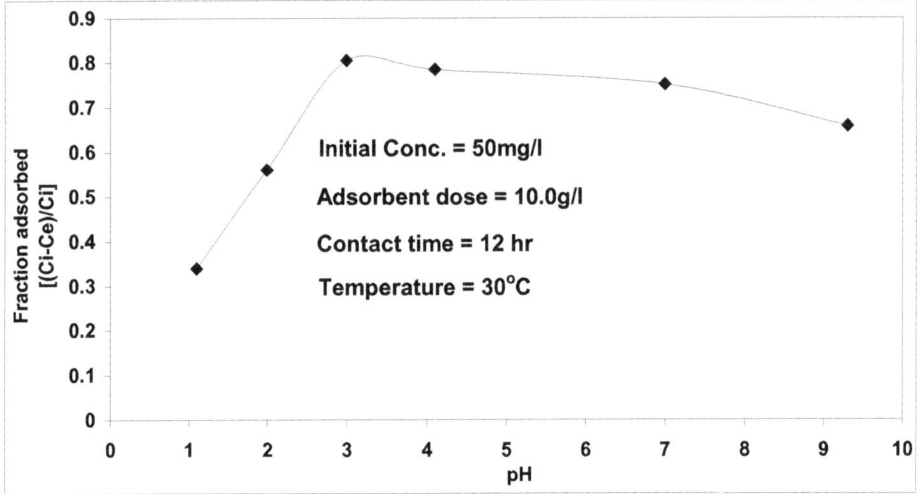

Figure 6.13: Effect of pH on Adsorption.

(c) Effect of Initial Cr (VI) Concentration on Adsorption

The effects of initial Cr (VI) concentrations on fraction adsorbed (Figure 6.14) were studied over the wide range of chromium concentration (5-100 mg/l). It may be observed that the chromium uptake is rapid during the initial period of adsorption and the maximum removal (100 - 93 per cent) is achieved at 5-20 mg/l concentration. The removal efficiency of chromium decreases when chromium concentration is increases. However the removal efficiency is recorded as 80.4 per cent at a concentration of 50 mg/l. In a similar study (Chand, *et al.*, 1994) obtained 90 per cent removal efficiency at a Cr (VI) concentration of 10 mg/l in a dose of 1.0 g/100 ml at a contact time of 1.5 hr when the pH of the solution was 2.0.

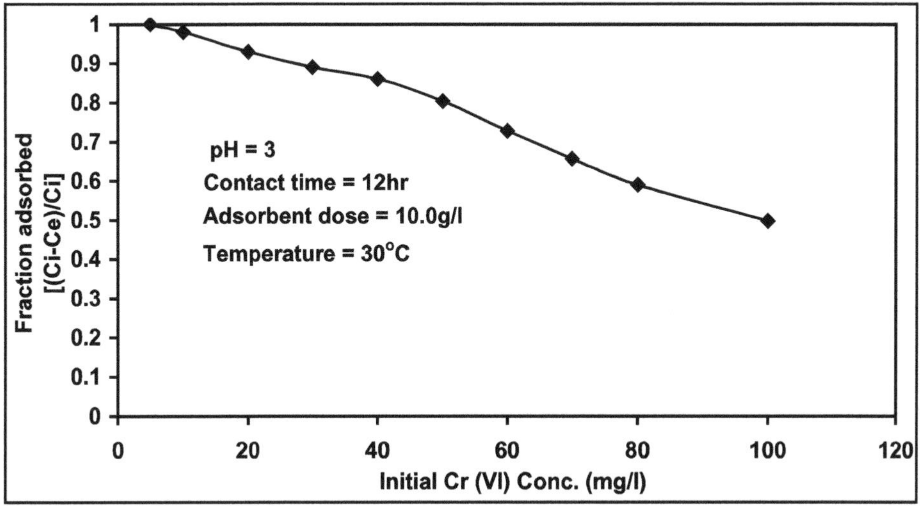

Figure 6.14: Effect of Cr (VI) Concentration on Adsorption.

(d) Effect of Contact Time and Particle Size on Adsorption

The adsorbent particle size has significant effect on the adsorption time and kinetics of adsorption. The influence of particle size furnishes important information for achieving optimum utilization of adsorbent and on the nature of breakthrough

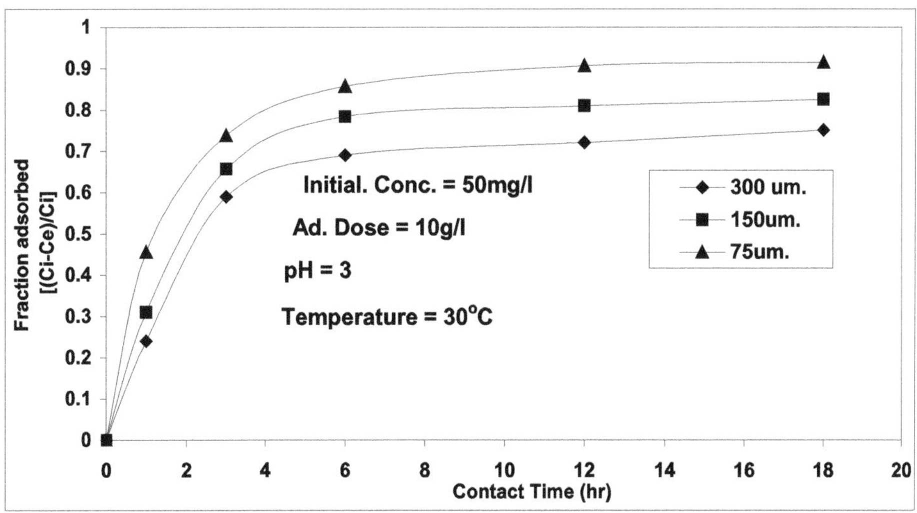

Figure 6.15: Effect of Contact Time and Particle Size on Adsorption.

curves for designing packed bed adsorbers. Three particle sizes 75 micron, 150 micron and 300 micron sieve (Indian standard Sieves) under optimal conditions of adsorbent dose, pH and contact time with an initial adsorbate concentration of 50 mg/l were studied. A plot of fraction adsorbed against contact time shown in Figure 6.15 indicates that, with an increase in adsorbent particle size, adsorption decreases. This may be explained on the basis of the surface area available for the adsorption, which is greater for small particle sizes (Poots and Healy, 1976). Similar behavior was found for the coconut shell.

(e) Effect of Contact Time and Temperature on Adsorption

The effect of temperature upon the adsorption rate was investigated at 30°C, 40°C and 50°C. The results are shown in Figure 6.16. It is observed that the mass of the chromium adsorbed per unit mass of adsorbent increases with the increasing temperature. This indicates the endothermic nature of the adsorption process. Several earlier investigators (Khare, *et al.*, 1987; Knocke and Hemphill, 1981; Panday, 1984; Panday, *et al.*, 1985, 1986; Singh, *et al.*, 1988) also have reported increase in uptake of adsorbate with increase in temperature.

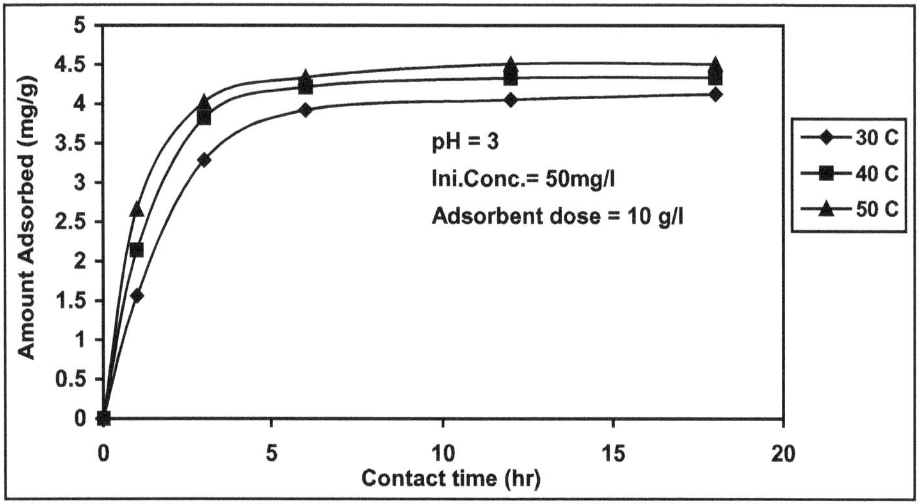

Figure 6.16: Effect of Contact Time and Temperature on Adsorption.

(f) Adsorption Isotherms

Similar to section 6.1.1 (f) R_l was determined by plotting 1/qe and 1/Ce. The curve for Langmuir and Freundlich adsorption isotherm is shown in Figure 6.17 and 6.18) and is found linear. The value of Langmuir isotherms and Freundlich isotherms, Coefficient of correlation Cc, Separation factor R_l and recommended isotherms were determined and shown in Table 6.3.

(g) Kinetics of Adsorption

The kinetic modeling for the removal of chromium (VI) by Neem bark has been carried out similar to section 6.1.1 (g). The straight-line plots of log (qe-q) verses t for

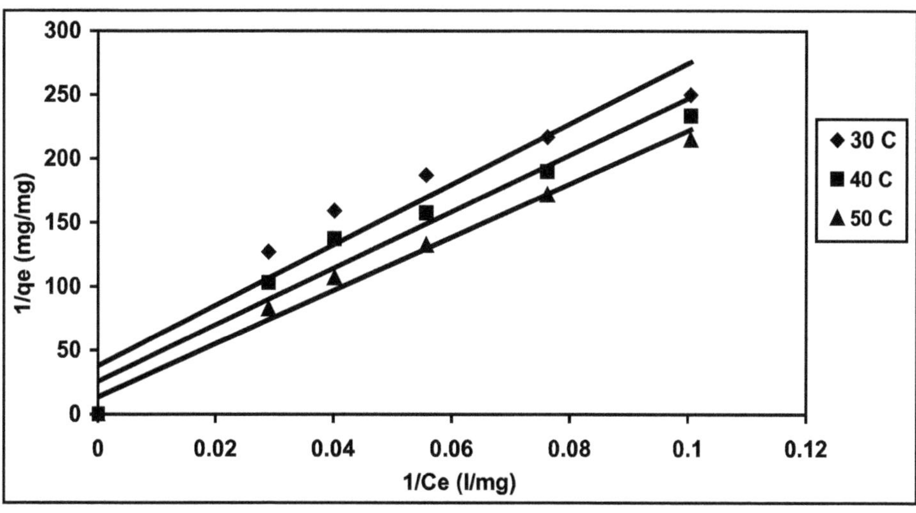

Figure 6.17: Langmuir Adsorption Isotherms Plot.

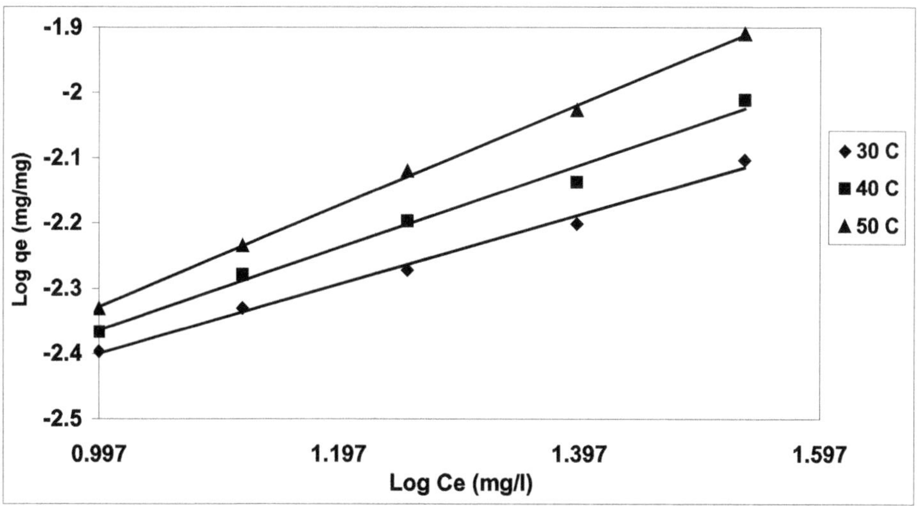

Figure 6.18: Freundlich Adsorption Isotherm Plot.

the adsorption show the validity of lagergren equation and suggest the first order kinetics (Figure 6.19). The rate constant K_{ad} is 0.3086. The value of E is evaluated from the calculated value of K_{ad} using equation 6.8 and found 1.55 kj/mole.

Table 6.3: Values of Langmuir and Freundlich Isotherms Constants

Temperature	*Langmuir Constants*				*Freundlich Constants*			*Recommended Isotherms*
	a	*b*	*Cc*	R_l	*K*	*1/n*	*Cc*	
30º C	2363.3	0.0160	0.909	0.555	.0011	0.530	0.992	$0.0011Ce^{0.530}$
40º C	2213.8	0.0115	0.949	0.634	.0010	0.631	0.987	$0.0010Ce^{0.631}$
50º C	2085.7	0.0063	0.985	0.760	.0008	0.776	0.998	$0.0008Ce^{0.776}$

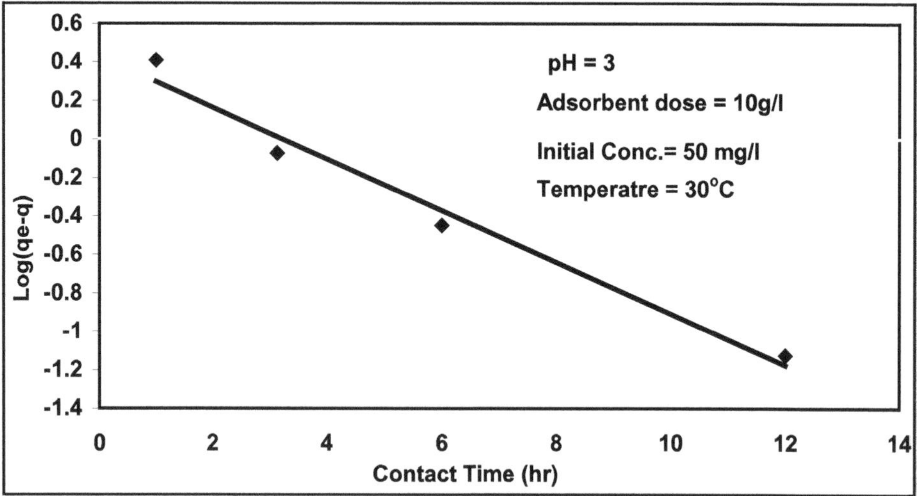

Figure 6.19: Lagergren Plot.

(h) Thermodynamic Parameters

The enthalpy change of sorption as calculated from the slope of ln b verses 1/T (Figure 6.20) is found to be 19.574 Kj/mole. The positive ΔH values confirm the endothermic nature of the sorption process and suggest the possibility of strong binding between sorbate and sorbent. The negative values of ΔG indicate the process to be feasible and spontaneous (Singh, *et al.*, 1982) and positive values of entropy suggested that randomness prevails at solid – solution interface due to adsorption of chromium on neem bark. The values of ΔG and ΔS at different temperature are listed in Table 6.4.

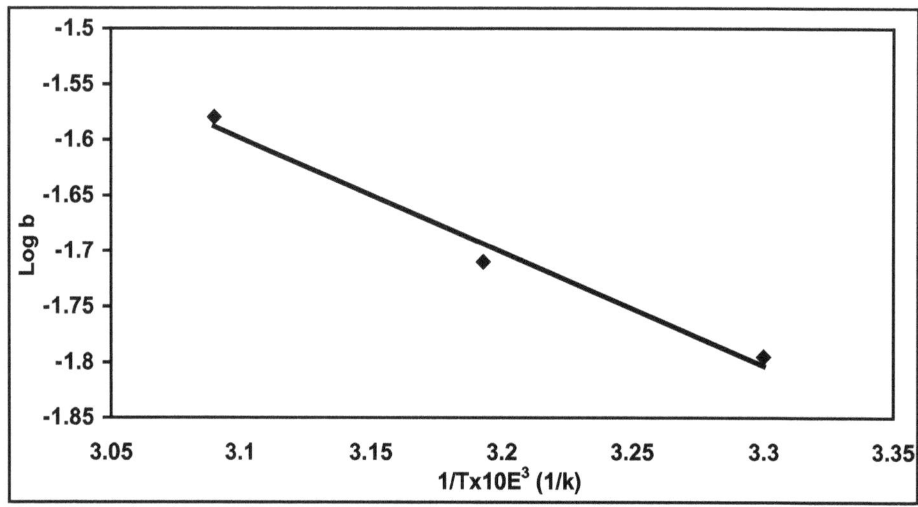

Figure 6.20: Van't Hoff Plot.

Table 6.4: Thermodynamic Parameters at Different Temperatures

Temperature (ºK)	*– ΔG, Kj/mole*	*ΔS, j/mole*
303	10.417	98.980
313	11.620	99.661
323	13.607	100.272

(i) Interparticle Diffusion Rate

Plot of amount adsorbed per unit mass of adsorbent versus square root of time is shown in Figure 6.21. For unsteady state adsorption a Ficks equation could be used to model on adsorption process and the quantity adsorbed on the solid surface will be proportional to the square root of the time (Bird, *et al.*, 1960). It is seen from this plot that there are three distinct linear sections, the initial steep linear portion (k_1) and the final relatively flat linear parts (k_2 and K_3). The initial linear part indicates that the interparticle diffusion, the later, less steeper linear, part suggested that adsorption is being controlled by the diffusion through its micro pores. The film or surface diffusion is an important rate controlling step and is a function of particle size, hydrodynamic conditions, physical properties, etc. in a well mixed adsorber. The adsorbate species are transported from the exterior to interior sites of the adsorbent particles and interparticle transport (within macro and micro-pores) is often the rate limiting step (Huang, *et al.*, 1984; Weber and Morris, 1963 a, b, 1964).

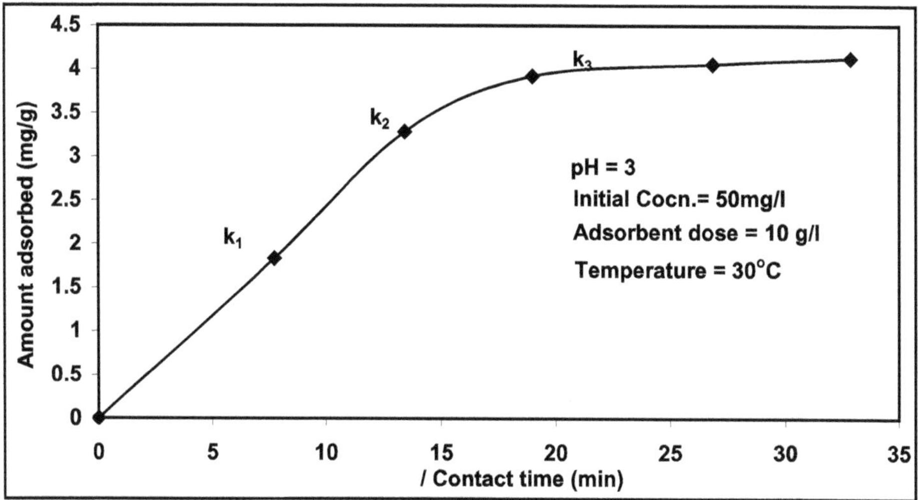

Figure 6.21: Interparticle Diffusion Plot.

6.2.2 Column Study

The breakthrough curve plotted in Figure 6.22. A flow rate of 1.0 l/d having metal concentration of 50mg/l (pH-3) was maintained. The column was run till the neem bark in the adsorption column gets exhausted and the treated effluents were analyzed at different time intervals. Column capacities were found greater than the batch capacities due to continuously large concentration at the interface of the sorption zone as the sorbate solution passes through the column while the concentration gradient decreases with time in a batch process (Srivastava, *et al.*, 1998).

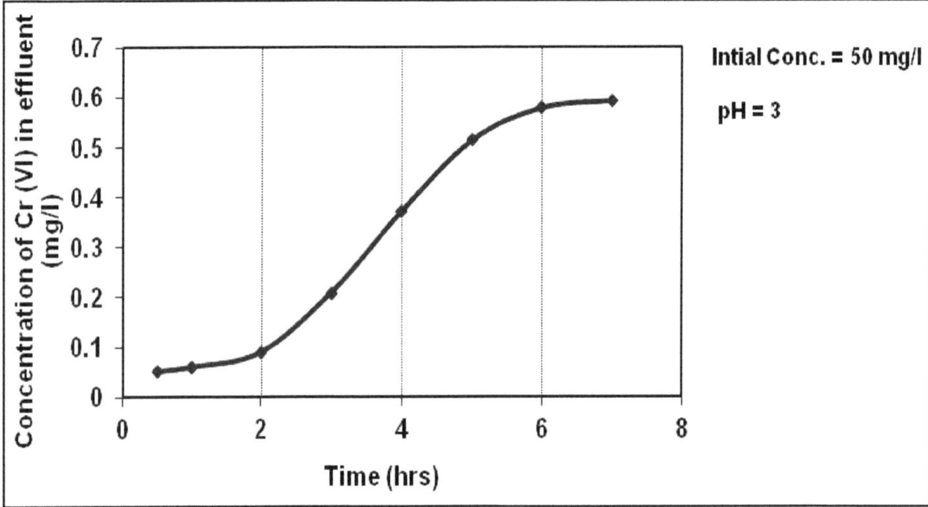

Figure 6.22: Breakthrough Curve.

6.2.3 Desorption and Hydrolysis Test

Desorption test was conducted for neem bark after their use in the equilibrium adsorption studies. About 10 g of saturated adsorbent was placed in a 300 ml capacity stoppered BOD bottle with distilled water and was shaken at room temperature for over two hours. After this the adsorbent was filtered and the suspension was analyzed for the chromium content. No chromium was detected in the water. Thus it indicates that adsorbed chromium was not being desorbed.

6.3 Sugarcane Bagasse

6.3.1 Batch Study Parameters

(a) Effect of Adsorbent Dose and Contact Time on Adsorption

The response of adsorbent dose and contact time on the removal of Cr (VI) is shown in Figure 6.23. Increase in the fractional adsorption occurs with corresponding increase in the dose of bagasse and contact time up to certain level, beyond which the fraction adsorbed remains constant. It is evident that a dose of 10g/l is sufficient to remove 65-90 per cent Cr (VI) in 0.5 –2.5 hours. The increase in the removal efficiency with simultaneous increase in adsorbent dose and contact time is due to the increase surface area and hence more active sites are available for the adsorption of Cr (VI).

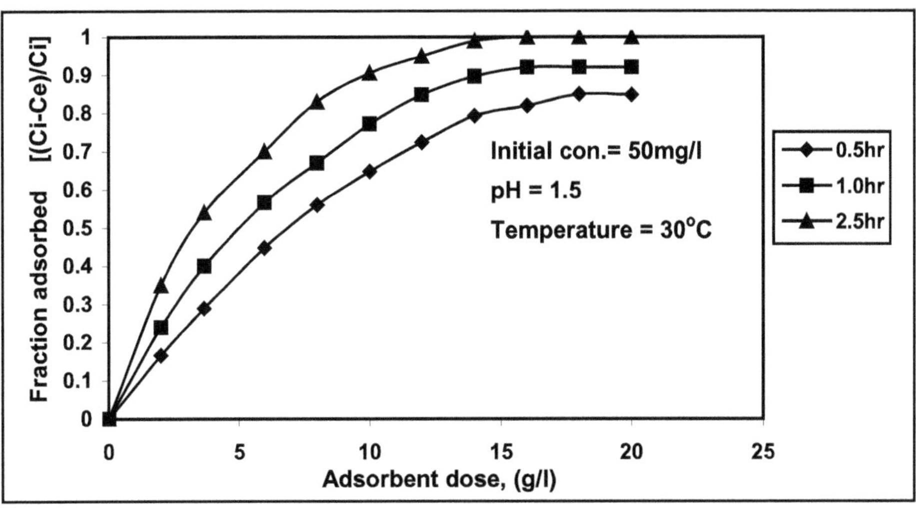

Figure 6.23: Effect of Adsorbent Dose and Contact Time on Adsorption.

(b) Effect of pH on Adsorption

Effect of pH on adsorption is shown in Figure 6.24. The results show the maximum removal is observed at pH 1.5 and on increasing the pH value it decreases. Approximately 4 percent removal is recorded at neutral pH 7. The reason for the better adsorption capacity observed at low pH values may be attributed to the large number of H^+ ions present at these pH values, which in term neutralize the negatively charged hydroxyl group (-OH) on adsorbed surface thereby reducing hindrance to the diffusion of dichromate ions. At higher pH, the reduction in adsorption may be possible due to abundance of OH^- ions causing increased hindrance to diffusion of positively charged dichromate ions. It is the common observation that the surface

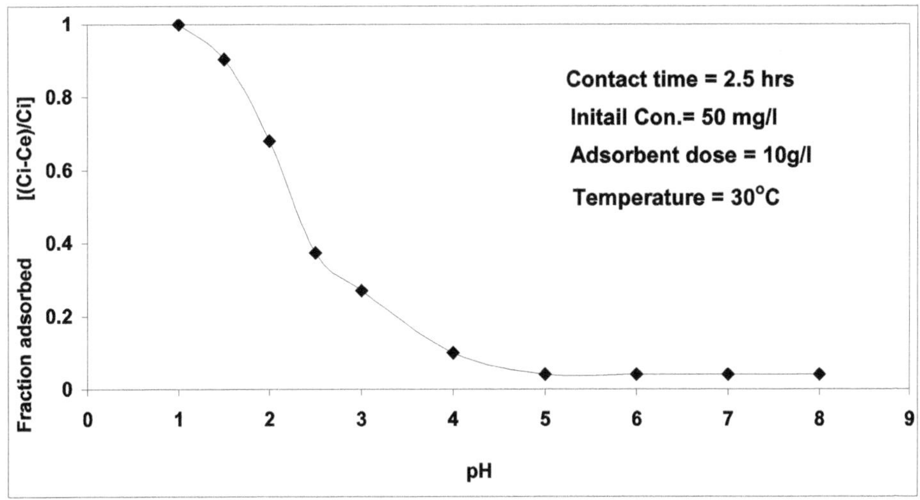

Figure 6.24: Effect of pH on Adsorbed.

adsorbs anions favorably in low pH range due the presence of H^+ ions (Gebehard and Coleman, 1974) whereas, the surface is active for the adsorption of cations at higher pH values due to the accumulation of OH^- ions (Huang and stunm, 1973).

(c) Effect of Initial Cr (VI) Concentration on Adsorption

The effects of initial Cr (VI) concentrations on fraction adsorbed (Figure 6.25) were studied over the wide range of chromium concentration (5-100 mg/l). It may be observed that the chromium uptake is rapid during the initial period of adsorption and the maximum removal (100- 96 per cent) is achieved at 5-20 mg/l concentration. The removal efficiency of chromium decreases when chromium concentration is increased. However the removal efficiency is recorded as 90.7 per cent at a concentration of 50 mg/l. In a similar study (Chand, *et al.*, 1994) obtained 90 per cent removal efficiency at a Cr (VI) concentration of 10 mg/l in a dose of 1.0 g/100 ml at a contact time of 1.5 hr when the pH of the solution was 2.0. In a significant deviation from the Neem and Coconut adsorption, a relatively higher adsorption is achieved even at higher concentration.

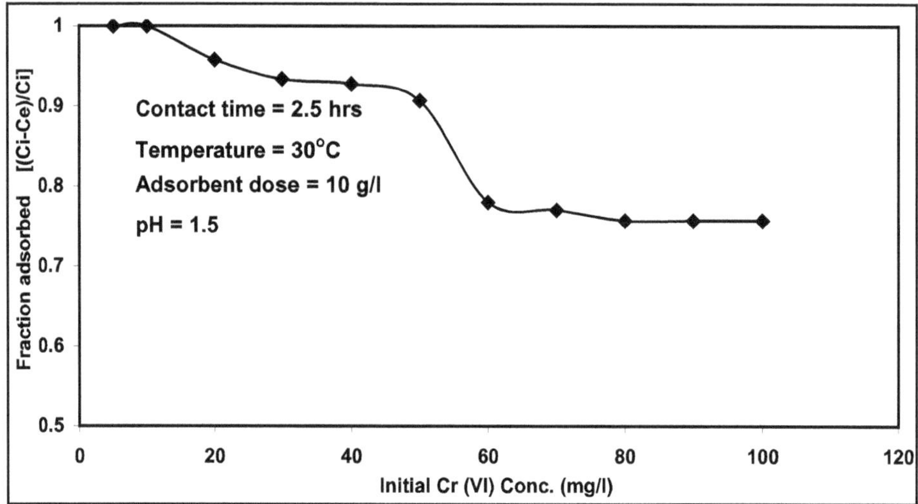

Figure 6.25: Effect Cr (VI) Concentration on Adsorption.

(d) Effect of Contact Time and Particles Size on Adsorption

The adsorbent particle size has significant effect on the adsorption time and kinetics of adsorption. The influence of particle size furnishes important information for achieving optimum utilization of adsorbent and on the nature of breakthrough curves for designing packed bed adsorbers. Three particle sizes 75 micron, 150 micron and 300 micron sieve (Indian standard Sieves) under optimal conditions of adsorbent dose, pH and contact time with an initial adsorbate concentration of 50 mg/l were studied. A plot of fraction adsorbed against contact time shown in Figure 6.26 indicates that, with increase in adsorbent particle size, adsorption decreases. This may be explained on the basis of the surface area available for the adsorption, which is greater for small particle sizes (Poots and Healy, 1976). Similar behaviors were found for the coconut shell and neem bark.

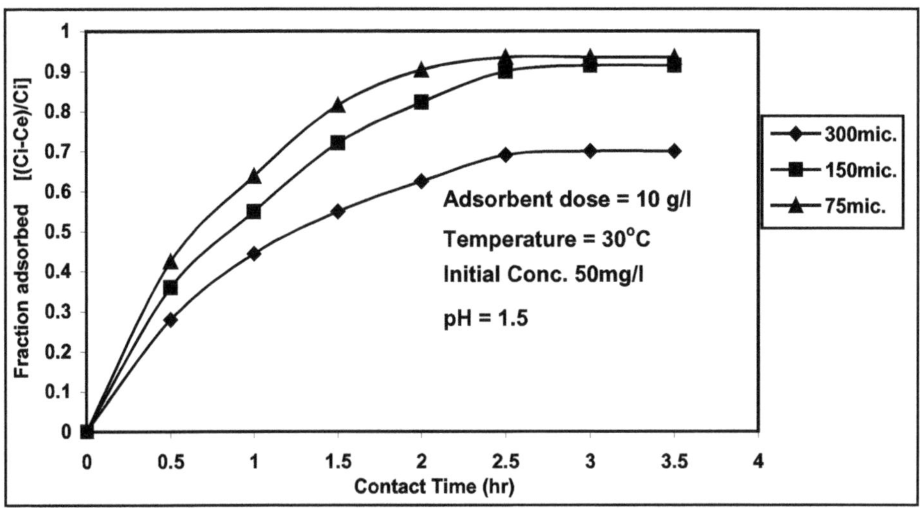

Figure 6.26: Effect of Contact Time and Particle Size on Adsorption.

(e) Effect of Contact Time and Temperature on Adsorption

The effect of temperature upon the adsorption rate was investigated at 30°C, 40°C and 50°C. The results are shown in Figure 6.27. It is observed that the mass of the chromium adsorbed per unit mass of adsorbent decreases with the increasing temperature. This indicates towards the exothermic nature of the adsorption process. Several earlier investigators (Cloutier, *et al.*, 1985; Gupta, *et al.*, 1991,1992; Johson, *et al.*, 1965; Mekey, *et al.*, 1980; Panday, *et al.*, 1984; Sharma, *et al.*, 1990 a, b, 1991)

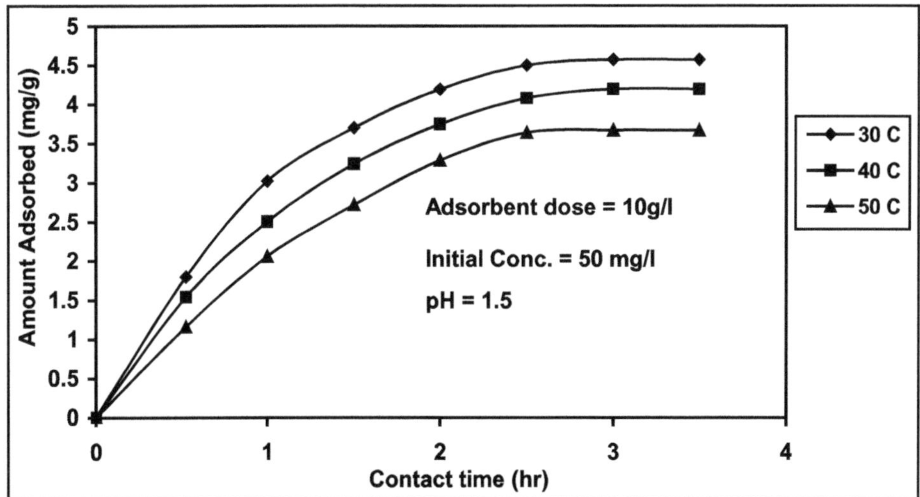

Figure 6.27: Effect of Contact Time and Temperature on Adsorption.

also have reported decrease in uptake of adsorbate with increase in temperature. However, it was observed that the chromium adsorbed per unit mass of the adsorbent increases with the increasing temperature in the cases of coconut shell and Neem bark.

(f) Adsorption Isotherms

Similar to section 6.1.1 (f) R_l was determined by plotting 1/qe and 1/Ce. The curve of Langmuir and Freundlich adsorption isotherms (Figure 6.28 and 6.29) are found linear respectively. The value of Langmuir isotherms and Freundlich isotherms, Coefficient of correlation Cc, Separation factor R_l and recommended isotherms were determined and are shown in Table 6.5.

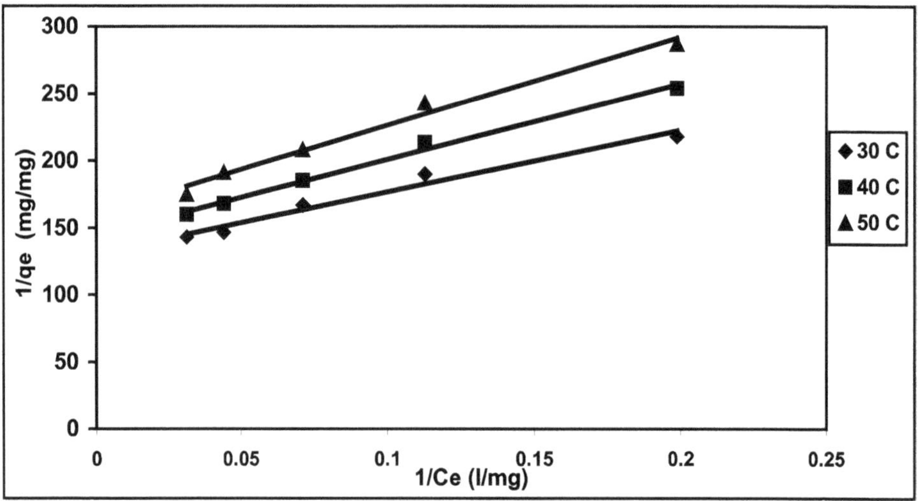

Figure 6.28: Langmuir Adsorption Isotherm Plot.

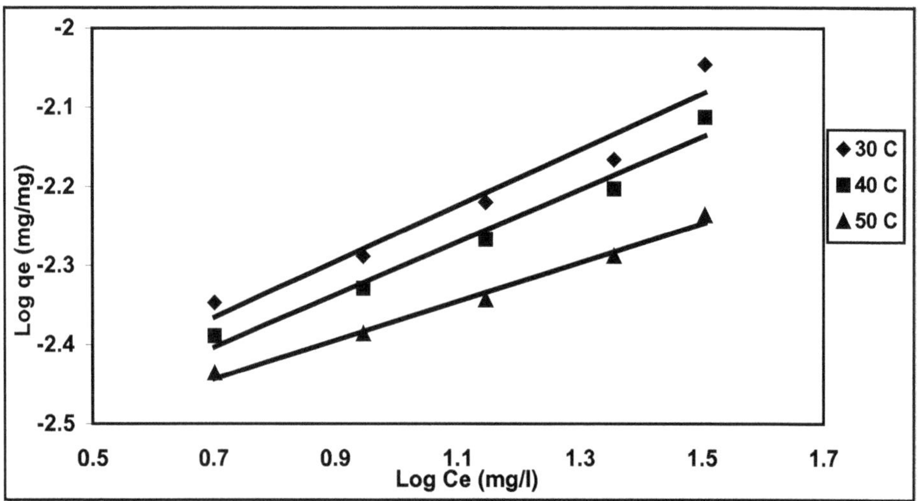

Figure 6.29: Freundlich Adsorption Isotherm Plot.

Table 6.5: Values of Langmuir and Freundlich Isotherms Constants

Temperature	*Langmuir Constants*				*Freundlich Constants*			*Recommended Isotherms*
	a	*b*	*Cc*	*R_l*	*K*	*1/n*	*Cc*	
30ºC	130.9	0.2855	0.973	0.065	0.0024	0.352	0.947	(130.90 x 0.2855Ce)/(1+130.9Ce)
40ºC	144.42	0.2558	0.993	0.072	0.0021	0.332	0.973	(144.42 x 0.2558Ce)/(1+144.42Ce)
50ºC	160.58	0.2440	0.984	0.075	0.0024	0.245	0.989	(160.58 x 0.2440Ce)/(1+160.58Ce)

(g) Kinetics of Adsorption

The kinetic modeling for the removal of chromium (VI) by sugarcane bagasse has been carried out similar to coconut shell and neem bark. The straight-line plots of log (qe-q) verses t for the adsorption show the validity of lagergren equation and suggest the first order kinetics (Figure 6.30). The rate constant K_{ad} is 0.365. The value of E is evaluated from the calculated value of K_{ad} using equation 6.8 and found 1.156 kj/mole.

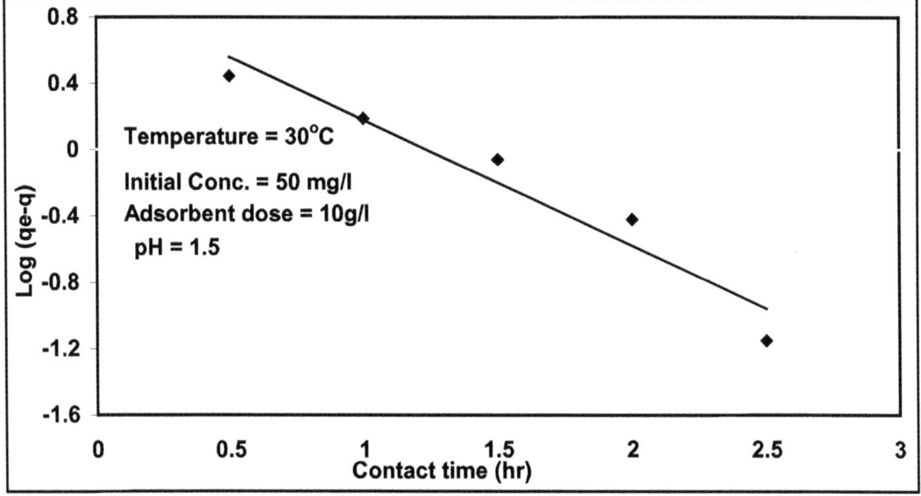

Figure 6.30: Lagergren Plot.

(h) Thermodynamic Parameters

The enthalpy change of sorption as calculated from the slope of ln b verses 1/T (Figure 6.31) is found to be –6.22 Kj/mole. The negative ΔH values confirm the exothermic nature of the sorption process. The negative values of ΔG indicate the process to be feasible and spontaneous (Singh, *et al.,* 1982) and negative values of entropy reflected the affinity of the adsorbent material. The values of ΔG and ΔS at different temperature are listed in Table 6.6. The negative values of free energy change for a system indicates spontaneity of adsorption process. Adsorption at

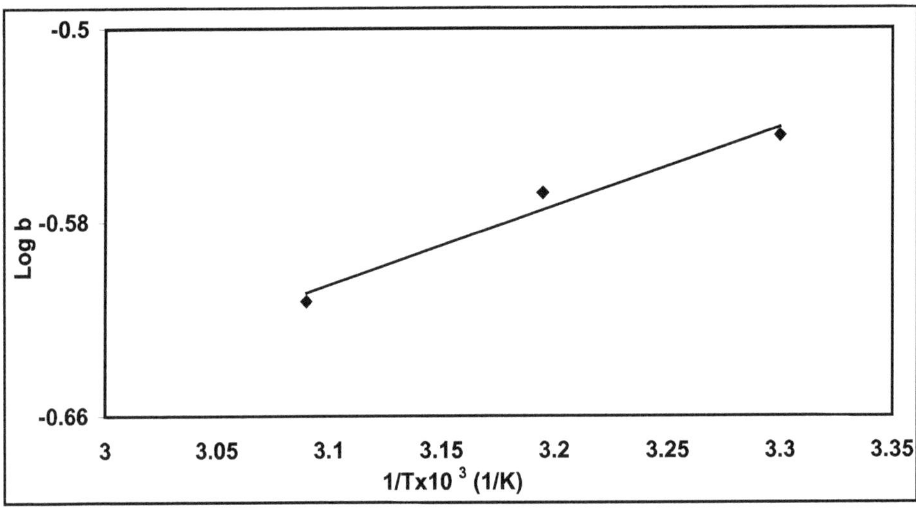

Figure 6.31: Van't Hoff Plot.

a solid solution interface generally shows an increase in entropy. This indicates, a faster interaction during the forward adsorption. Association, fixation or immobilizations of adsorbate on the interface between two phases result in loss of the degree of freedom there by, showing a negative entropy effect. The negative values of entropy have been reported earlier in the case of the adsorption process. Cr (VI) on flyash – wallastonite (Panday, *et al.*, 1984). Cr (VI) on blast furnace flue dust and COD (chemical oxygen demand) on fly ash (Baisakh, *et al.*, 1996). Adsorption on coal char (Baisakh, *et al.*, 2002).

Table 6.6: Thermodynamic Parameters at Different Temperatures

Temperature (ºK)	*– ΔG, Kj/mole*	*ΔS, j/mole*
303	3.15	10.132
313	3.55	8.530
323	3.78	7.529

(i) Interparticle Diffusion Rate

For unsteady state adsorption a Ficks equation could be used to model on adsorption process and the quantity adsorbed on the solid surface will be proportional to the square root of the time (Bird, *et al.*, 1960). The transport of adsorbate from solution phase to the surface of adsorbent particle is controlled either by film diffusion, pore diffusion, pore surface diffusion and adsorption on the pore wall or by the combined effects of more then one of these factors (Crank, 1956; Keinath, 1977; Weber, 1972). Plot of amount adsorbed per unit mass of adsorbent versus square root of time is shown in Figure 6.32. It is seen from this plot that there are three distinct linear sections, the initial steep linear portion (k_1) and the final relatively flat linear parts (k_2 and k_3. The initial linear part indicates that the interparticle diffusion, the latter, less steeper linear, parts suggested that adsorption is being controlled by the micro pores.

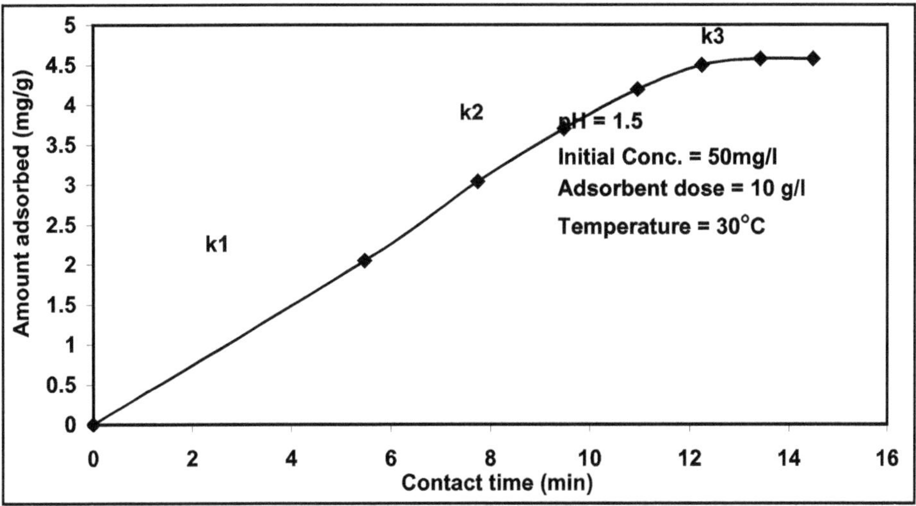

Figure 6.32: Interparticle Diffusion Plot.

6.3.2 Column Study

The breakthrough curve plotted in Figure 6.33. A flow rate of 1.0 l/d having metal concentration of 50mg/l (pH-1.5) was maintained. The column was run till the bagasse in the adsorption column gets exhausted and the treated effluent was analyzed at different time intervals. Column capacities were found greater

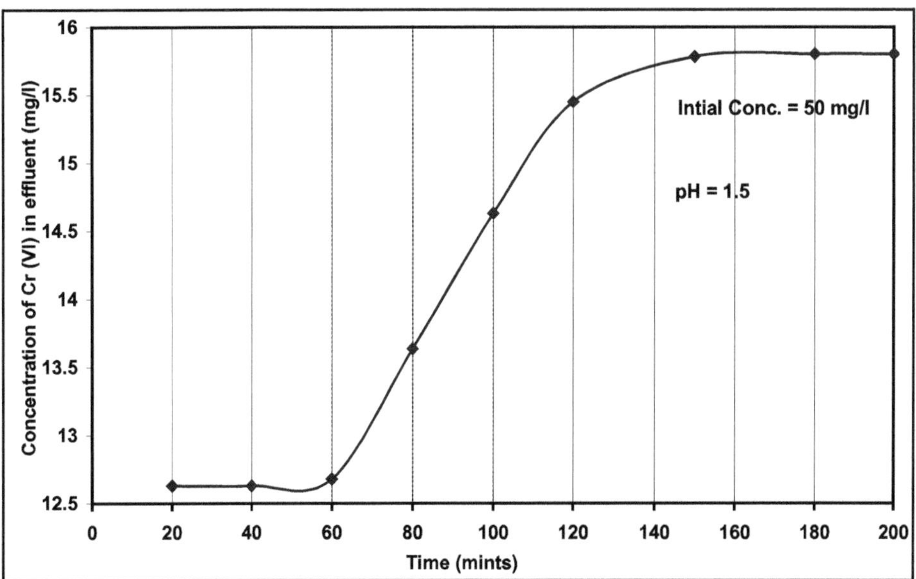

Figure 6.33: Breakthrough Curve.

than the batch capacities due to continuously large concentration at the interface of the sorption zone as the sorbate solution passes through the column while the concentration gradient decreases with time in a batch process (Srivastava *et al.*, 1998).

6.3.3 Desorption and Hydrolysis Test

Desorption test was conducted for bagasse after their use in the equilibrium adsorption studies. About 10 g of saturated adsorbent was placed in a 300 ml capacity stoppered BOD bottle with distilled water and was shaken by shaker at room temperature for over two hours. After this the adsorbent was filtered and the suspension was analyzed for the chromium content. No chromium was detected in the water. Thus it indicates that adsorbed chromium was not being desorbed.

6.4 Scanning Electron Microscopy

Electron microscopy, in principle, is similar to the optical microscope except for the enhancement of the resolution power enabling electron microscopy to examine the minor details of the objects. Electron microscopy can be classified into two groups; transmission electron microscopy (TEM) and scanning electron microscopy (SEM). In transmission electron microscopy, properly prepared samples are replicas are placed in the electron beam. Electrons pass through the surface and are collected on screen. In the case of scanning electron microscopy, a beam of electrons is bombarded on a specially prepared sample surface. The electrons emitted from this sample are arbitrarily divided into secondary electros, back scattered electrons, or x-ray photons and collected in the detector. Later, these signals are amplified and displayed on the screen. TEM has a lower resolution limit of less than 5 A (magnification up to 300,000) compared to 250 A for scanning microscopy.

Electron Microscope is form of a microscope that uses a beam of electrons instead of a beam of light (as in the optical microscope) to form a large image of a very small object. In optical microscopes the resolution is limited by the wavelength of the light. High-energy electrons, however, can be associated with a considerably shorter wavelength than light; for example, electrons accelerated to an energy of 10^5 electron volts have a wavelength of 0.04 nanometer enabling a resolution of 0.2 - 0.5 nm to be achieved. The transmission electron microscope has an electron beam, sharply focused by electron lenses, passing through a very thin metallized specimen (less than 50 nanometers thick) onto a fluorescent screen, where a visual image is formed. This image can be photographed. The scanning electron microscope can be used with thicker specimens and forms a perspective images, although the resolution and magnification are lower. In this type of instrument a beam of primary electrons scans the specimen and those that are reflected, together with any secondary electrons emitted, are collected. This current is used to modulate a separate electron beam in a TV monitor, which scans the screen at the same frequency, consequently building up a picture of the specimen. The resolution is limited to about 10-20 nm.

In the present investigation, scanning electron microscopy (SEM) was used to study surface characteristics of the adsorbents before and after adsorption. The scanning electron microscopy of all the adsorbents was conducted in All India

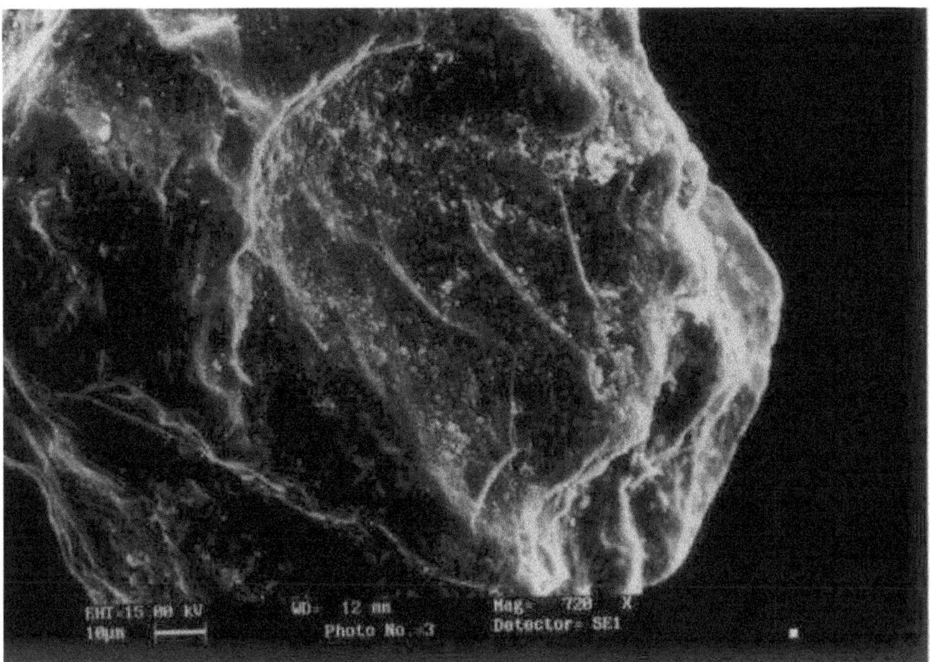

Figure 6.34: Scanning Electron Microscopy of the Adsorbent (Coconut Shell) before Adsorption.

Figure 6.35: Scanning Electron Microscopy of the Adsorbent (Coconut Shell) after Adsorption.

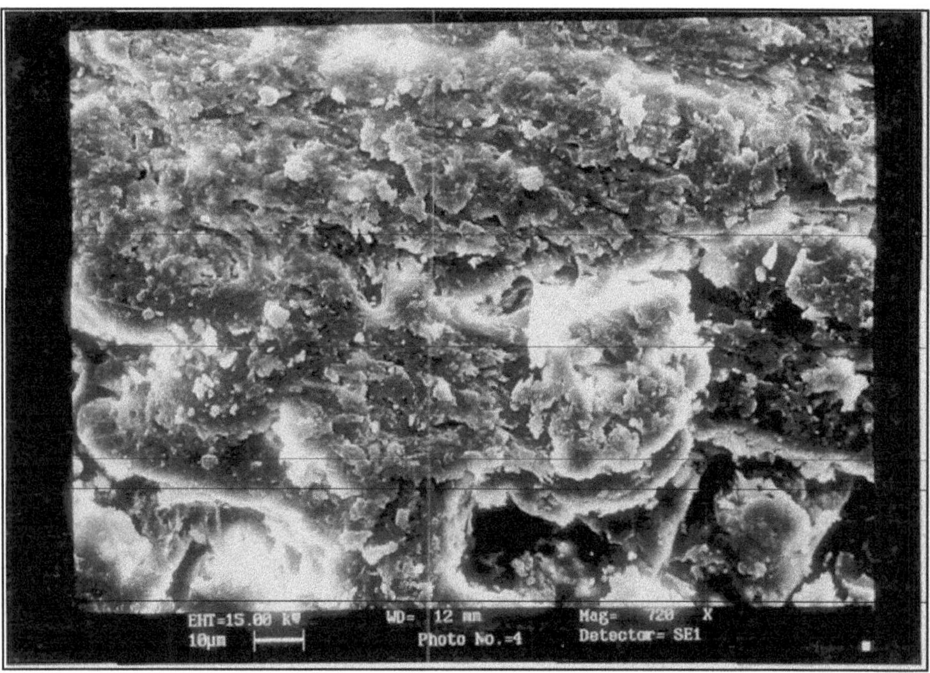

Figure 6.36: Scanning Electron Microscopy of the Adsorbent (Neem Bark) before Adsorption.

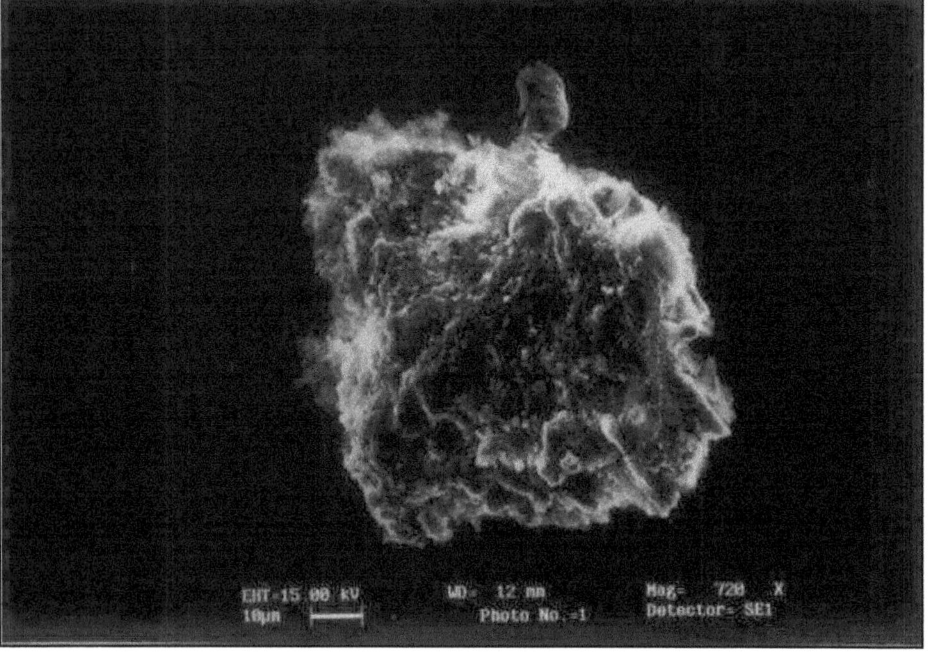

Figure 6.37: Scanning Electron Microscopy of the Adsorbent (Neem Bark) after Adsorption.

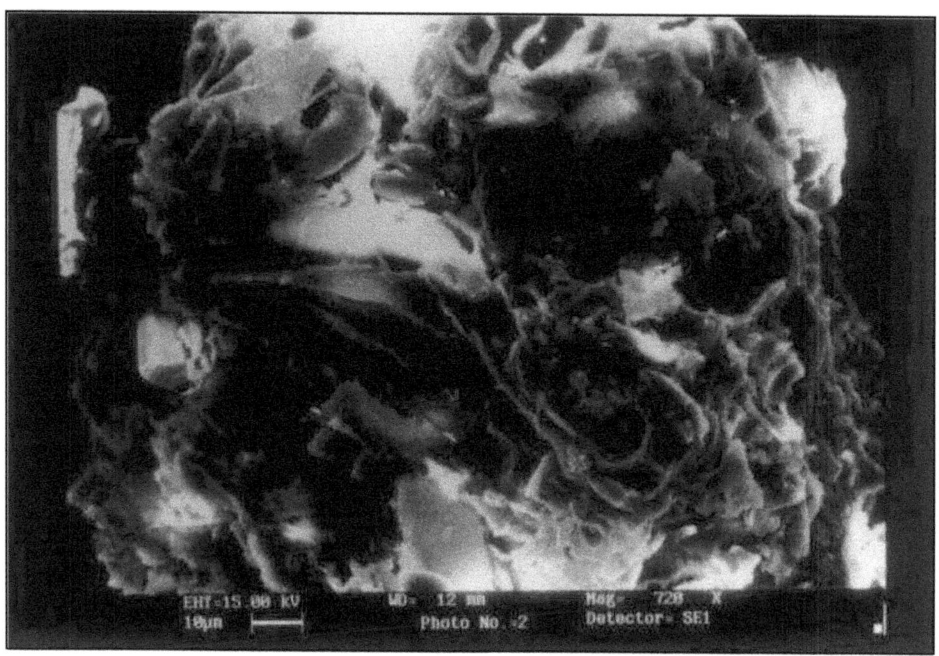

Figure 6.38: Scanning Electron Microscopy of the Adsorbent (Raw Bagasse) before Adsorption.

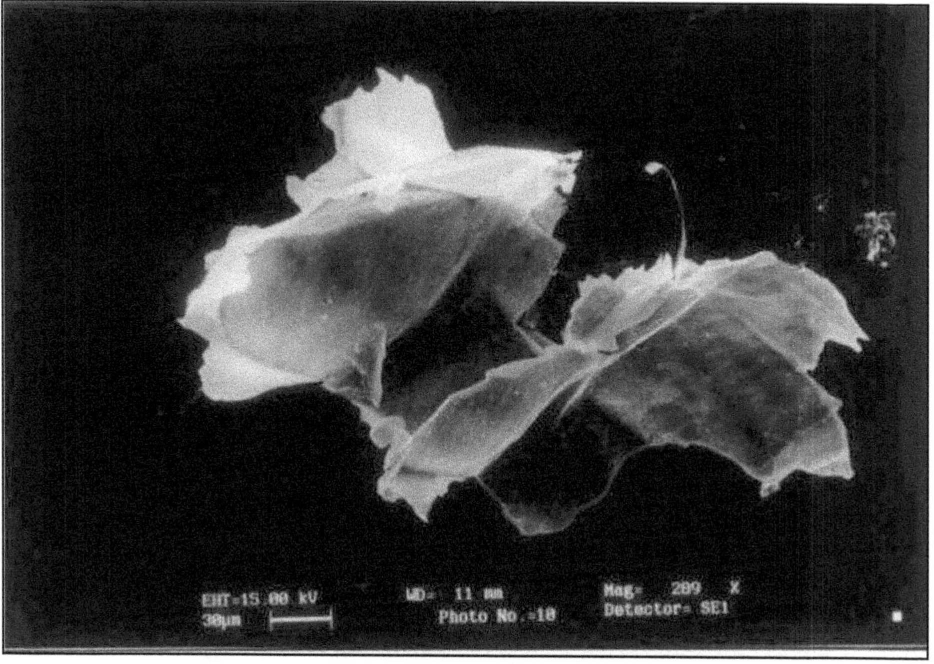

Figure 6.39: Scanning Electron Microscopy of the Adsorbent (Raw Bagasse) after Adsorption.

Institute of Medical Sciences (AIMS) central facility laboratory, New Delhi (India). Figures 6.34 - 6.39 show typical photographs for Coconut shell, Neem bark and Sugarcane bagasse before and after adsorption. From these figures, it is evident that the structure of the above adsorbents has high percentage of pores before the adsorption, which would be suitable for adsorption of chromium ions. A large surface area is also evident from these photographs, which ultimately enhance the adsorptive capacity. From the SEM photographs it is also evident that the honeycombing type porous structure found in the adsorbent surface is collapses to some extent. This may be due to the fact that adsorbent being a natural biological material changes its characteristics.

6.5 Comparison of Coconut Shell, Neem Bark and Sugarcane Bagasse Adsorbents

The investigations reveal that agricultural waste adsorbents surfaces (Coconut shell, Neem bark and Sugarcane bagasse) can be used as effective adsorbents for the removal of hexavalent chromium from electroplating and metal finishing wastewaters. Sorption of chromium is highly pH dependent as shown in Figure 6.40 and the best results were obtained in the pH range 1.5-3.0. The reason for the better adsorption capacity observed at low pH values may be attributed to the large number of H^+ ions present at these pH values, which in term neutralize the negatively charged hydroxyl group (-OH) on adsorbed surface thereby reducing hindrance to the diffusion of dichromate ions. At higher pH, the reduction in adsorption may be possible due to abundance of OH^- ions causing increased hindrance to diffusion of positively charged dichromate ions. It is the common observation that the surface adsorbs anions favorably in low pH range due the presence of H^+ ions (Gebehard and Coleman, 1974) whereas, the surface is active for the adsorption of cations at higher pH values due to the accumulation of OH^- ions (Huang and stunm, 1973).

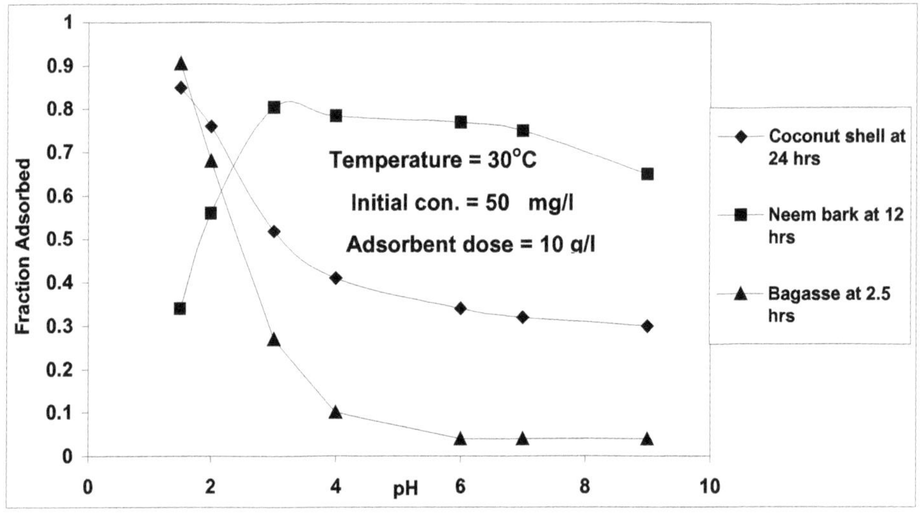

Figure 6.40: Effect of pH on Adsorption.

Similar observations were recorded during the studies conducted on dihydric phenol removal on activated carbon by Mahesh, *et al.* (1999) and Sharma, *et al.* (1993) used sphagnum moss peat for the removal of chromium.

It is has been observed that the percentage removal increases with increasing adsorbent dose. 10 g/l dose of sugarcane bagasse is sufficient to adsorb more than 90 per cent Cr (VI) having 50 mg/l initial concentration within 1.0 hour. On further increasing the adsorbent dose to 16 g/l a 100 per cent removal efficiency were observed as shown in Figure s6.41. Similarly 10 g/l adsorbent dose of neem bark adsorbs more than 80 per cent at pH 3.0 and coconut shell 85 per cent at pH 1.5 respectively. Similar results were reported by (Ayub, *et al.*, 2001, 2002; Rao, *et al.*, 2002; Bansal and Sharma, 1992; Kim and Joltech, 1977; Singh, *et al.*, 1992; Mall, 1992).

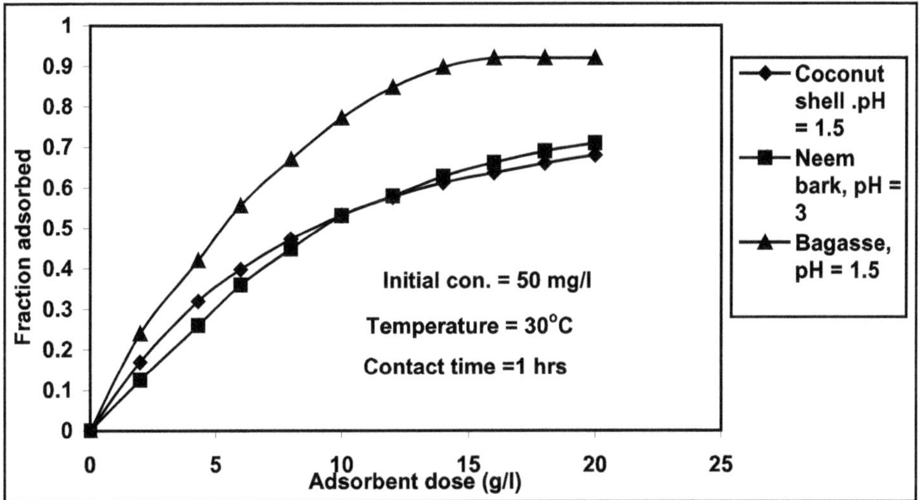

Figure 6.41: Effect of Adsorbent Dose on Adsorption.

At the initial stage, Cr (VI) removal is very fast and the later stage of Cr (VI) removal is much slower. The extent of removal depends on the metal concentration, adsorbent dose, contact time, particle size and pH. Several investigators have reported that appreciable reduction in percent adsorption with increase in metal ion concentration (Mckay, *et al.*, 1991; Panday, 1984; Kumar, 1987). Sugarcane bagasse is found to be superior to neem bark and coconut shell, as shown in Figure 6.42 with respect to the removal efficiency of the above metal in 10 g/l adsorbent dose and having 50 mg/l initial concentration of the solution. The removal efficiency ranges from 80 per cent - 90 per cent.

For all the adsorbents, the Langmuir and Freundlich adsorption isotherms were plotted and the isotherms are given in the Table 6.7. It is observed that best fit straight line having correlation coefficient (Cc) value of 0.917 and 0.920 for coconut shell, 0.909 and 0.992 for Neem bark and 0.973 and 0.947 for bagasse. Hence it can be stated that the adsorption isotherm is represented more closely by Freundlich adsorption isotherm for coconut shell and neem bark, while Langmuir adsorption isotherm for Bagasse.

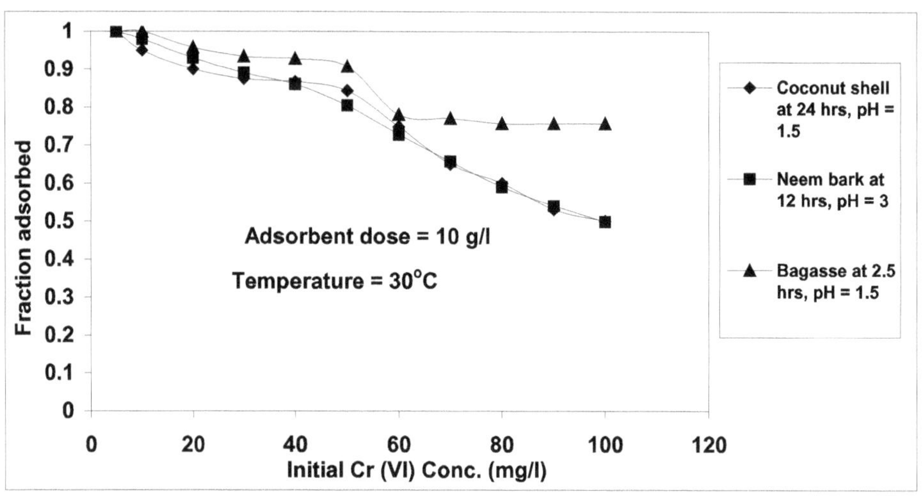

Figure 6.42: Effect of Initial Cr (VI) Concentration on Adsorption.

Table 6.7 Values of Langmuir and Freundlich Isotherms Constants at 30º C

Adsorbent	*Langmuir Constants*			*Freundlich Constants*			*Recommended Isotherm*
	a	*b*	*Cc*	*K*	*1/n*	*Cc*	
Coconut shell	1900.9	0.0220	0.917	0.0010	0.555	0.920	$0.0010Ce^{0.555}$
Neem bark	2363.3	0.0160	0.909	.0011	0.530	0.992	$0.0011Ce^{0.530}$
Bagasse	130.9	0.2855	0.973	0.0024	0.352	0.947	(130.6x 0.2855Ce) (1+130.6Ce)

The change in apparent enthalpy (ΔH), free energy (ΔG) and entropy (ΔS) of sorption were calculated using thermodynamic equations and values at 30 °C are listed in Table 6.8. The positive ΔH values confirm the endothermic nature of the sorption process and suggested the possibility of strong binding between sorbate and sorbent for coconut shell and Neem bark. While negative ΔH values confirm exothermic nature of adsorbent. Negative values of ΔG indicate the process to be feasible and spontaneous. The negative values of free energy change for a system indicates spontaneity of adsorption process. Adsorption at a solid solution interface generally shows an increase in entropy. This indicates, a faster interaction during the forward adsorption. Association, fixation or immobilizations of adsorbate on the interface between two phases result in loss of the degree of freedom there by, showing a negative entropy effect. The positive entropy in case of Coconut shell and Neem bark may be attributed to an increase in translational entropy caused by the randomness of displaced water molecules from the surface of the adsorbent (Wright and Pratt, 1974). The negative values of entropy have been reported earlier in the case of the adsorption process. Cr (VI) on fly ash – wallastonite (Panday, *et al.*, 1984). Cr (VI) on blast furnace flue dust and COD (chemical oxygen demand) on fly ash (Baisakh, *et al.*, 1996). Adsorption on coal char (Baisakh, *et al.*, 2002).

Table 6.8: Thermodynamic Parameters at 30ºC

Adsorbent	*– ΔH, Kj/mole*	*– ΔG, Kj/mole*	*ΔS, j/mole*
Coconut shell	14.23	9.614	78.693
Neem bark	19.574	10.417	98.980
Bagasse	–6.22	3.150	–10.132

Chapter 7

Conclusion and Recommendations

The following conclusions could be drawn on the basis of the present study

1. The investigations show that agricultural waste adsorbents (Coconut shell, Neem bark and Sugarcane bagasse) can be used as effective adsorbents for the removal of hexavalent chromium from electroplating and metal finishing wastewaters. It is concluded that adsorption behavior is dependent upon the nature of adsorbent and all the agro-based materials would not behavior identical.
2. It is seen that the percentage removal increases with increasing adsorbent dose. 10 g/l dose of sugarcane bagasse is sufficient to adsorb more than 90 per cent Cr (VI) having 50 mg/l initial concentration within 1.0 hour at pH 1.5. On further increasing the adsorbent dose to 16 g/l, 100 per cent removal efficiency were observed. Similarly 10 g/l adsorbent dose of neem bark adsorbs more than 80 per cent at pH 3.0 and coconut shell 85 per cent at pH 1.5 respectively.
3. The chromium uptake capacity of all the adsorbents investigated is found highly pH dependent and the best results were obtained in the pH range 1.5-3.0.
4. The initial stage of Cr (VI) removal is found very fast and the later stage of Cr (VI) removal is much slower up to saturation limit. Percentage removal of chromium increases with decrease in metal concentration. The extant of removal depends on the metal concentration, adsorbent dose, contact time, particle size and pH. Sugarcane bagasse is found to be superior to neem bark and coconut shell with respect to the removal efficiency.

The removal efficiency ranges from 80 per cent - 90 per cent for all the adsorbents investigated.

5. Chromium adsorption follows first order rate equation for all the adsorbents studied.
6. The isotherm data obtained is more closely to Freundlich adsorption isotherm for coconut shell and neem bark. Sugarcane bagasse follows Langmuir adsorption isotherm.
7. The positive ΔH values confirm the endothermic nature of the sorption process and suggested the possibility of strong binding between sorbate and sorbent for coconut shell and Neem bark. While negative ΔH values confirm exothermic nature of adsorbent. Negative values of ΔG indicate the process to be feasible and spontaneous. Positive values of ΔS reflect the affinity of the adsorbent material.
8. Plot of amount adsorbed per unit mass of adsorbent versus square root of time shows the initial linear part interparticle diffusion, latter, less steeper linear, part suggested that adsorption is being controlled by the micro pores.
9. Desorption/leaching test was conducted for all the adsorbents after their use in the equilibrium adsorption studies. No chromium was detected in the water. Thus it indicates that adsorbed chromium was not being desorbed.
10. The agro-waste adsorbents studied do not hydrolysis and BOD of leachete does not change. It means that the adsorbents are stable materials under the process conditions.
11. The exhaustive capacity of the sorbent is higher in case of column process than the batch process. This is due to continuously large concentration at the interface of the sorption zone as sorbate passes through the column.
12. The agricultural waste adsorbents may be found superior than activated carbon. Since they are relatively cheaper and locally available. Waste disposal is easier in case of rural appropriate technology is evolved. These adsorbents require only alkali/acid treatment to increase their efficiency.
13. Regeneration studies are not necessary with the view that the cost of the adsorbents is very low and it can be disposed off safely or burned after drying. Although safe disposal of adsorbent need to be discussed and investigated.
14. It is proposed that pilot plant studies should be conducted in order to scale-up the process and determine the realistic cost benefit analysis. The utilization of agro-waste materials for the removal of chromium could be very beneficial in the development of rural friendly appropriate technologies. It could reduce the probability of heavy metal contamination in water supply in the rural areas.

References

1. Aggarwal, D., Goyal, M., and Bansal, R. C. (1999). "Adsorption of chromium by activated carbon from aqueous solution". *Carbon*, 37, 12, 1989-97.
2. Agro, D. G., and Culp G. L. (1972). *Wat. Sew. Wks.*, 119,9, 128.
3. Ahmad, S. M. (1972). " Electrical double layer at metal oxide- solution interface in oxide- solution interface in oxide and oxide films". *Diggle, J. W. Ed., Marcel dekker, New York.*
4. Ajmal, M. (1993). "Surface Entrapment of toxic metals from electroplating waste and their possible recovery ". *Water, Air and Soil Pollution*, 68, 485 – 492.
5. Ajmal, M., Mohd, A., Yousuf, R., and Ahmad, A. (1998). "Adsorption behavior of cadmium, zinc, nickel and lead from aqueous solutions by *Mangifera indica* seed shell". *Indian Jour. of Envion. Hlth*, 40, 01, 15.
6. Ajmal, M., Rao, R. A. K., Ahmad, R., and Ahmad J. (2000). "Adsorption studies on *Citrus reticulata* (fruit peel of orange): removal and recovery of Ni (II) from electroplating wastewater". *Journal of Hazd. Materials*, 49, 1-2, 117-131.
7. Ajmal, M., Rifaqat, A. K. R., and Bilquees, A. S. (1995). "Adsorption and removal of dissolved metals using pyrolusite as adsorbent". *Environmental Monitoring and Assessment*, 38, 25-35.
8. Alaerts, G. J., Jitjatuarant, V., and Kelderman, P. (1989). "Use of coconut shell based activated carbon for chromium (VI) removal". *Wat. Sci. Tech.*, 21, 1701. (Quoted by Ramos, *et al.*, 1994).
9. Ali, M., and Namita, D. (1992). "Effect of pH on Adsorption process of chromium (VI) with a new low cost adsorbent". *Indian J. of Environ. Pollution.* 12, 3, 202-209.
10. Ananthakrishna, N. P., and Parvathy, B. (1982). "Adsorption of Cr (VI) from aqueous solution by bottom ash from thermal pollution". *Journal of Indian Association for Water Pollution Control Tech.* Annual, IX, 176-181.

11. Anonymous. (1971). "An Investigation of techniques for removal of chromium from electroplating wastes". *US EPA publication*, 12010EIE (Quoted by Patterson and Minear, 1985).

12. Anonymous. (1973). " Development Document for proposed effluent limitations guidelines and new source performance standards- Copper, Nickel, chromium and Zinc segment of the electroplating point source category". *US Environmental Protection Agency document*. EPA – 440/1-73- 003 (Quoted by Patterson and Minear, 1985).

13. Ansari, M. H., Deshkar, A. M., Dharmadhikari, D. M., Pentu S. S., and Hasan M. Z. (1999). " Neem (Azadorachta indica) bark for the removal of mercury from water". Conference at Nagpur, India.

14. Ariyapadi, S. V., Ramesh, K. K. T., and Rohani, S. (1999). "A note on hexavalent chromium removal using chelating ion- exchange resin". *Jour. of Indian Chem. Engineer*, Section B, 41:04 CHE, 138-140.

15. Arulnatham, A., Balasubramain, N., and Ramakrishna, T.V.(1989). " Coconut shell for treatment of cadmium and lead containing wastewater ". *Metal finishing*, 87, 51-55.

16. Ayub, S., Ali, S. I., and Khan, N. A (2001). " Efficiency Evaluation of Neem bark (*Azadirachta indica*) bark in the treatment of industrial wastewater ". *Environmental Pollution Cont. Journal*, vol.4 (4), 34-38.

17. Ayub, S., Ali, S. I., and Khan, N. A. (1998). "Treatment of Wastewater by Agricultural Wastes". *Environmental Pollution Cont. Journal*, 2 (1).

18. Ayub, S., Ali, S. I., and Khan, N. A. (1999). "Extraction of Chromium from the wastewater by Adsorption". *Environmental Pollution Cont. Journal*. 2, (5).

19. Ayub, S., Ali, S. I., and Khan, N. A. (2002). "Adsorption studies on the low cost adsorbent for the removal of Cr (VI) from the electroplating industries". *Environmental Pollution Cont. Journal*, 5 (6).

20. Baes, A. U., Okuda, T., Nishijima, Wataru., Shoto, E., and Okada, M. (1997). "Adsorption and ion exchange of some groundwater anion contaminants in an amine modified coconut coir". *Wat. Sci. Tech.*, 35,7, 89-95.

21. Baily, S. E., Olin, T. J., Bricka, R. M., and Adrian, D. D. (1999). " A review of potentially low- cost sorbents for heavy metals ". *Water Research*, 33 (1), 2469-2479.

22. Baisakh, P. C., Patnaik, S. N. (2002). " Removal of Hexavalent chromium from aqous solution by adsorption on coal char". *Ind. J. of Env. Hlth*, 189-196.

23. Baisakh, P. C., Patnaik, S. N., and Patnaik, L. N. (1996). "Removal of COD from textile mill effluent using flyash". *Ind. J. Env. Protection*, 16 (2), 135-139.

24. Bandyyopaddyay, A., and Biswas, M. N. (1998). " Removal of hexavalent chromium by synergism modified adsorption". *Indian Journal of Env. Pollution*, 18, 09, 662-671.

25. Bansal, T. K., and Sharma, H. R. (1992). "Chromium removal by adsorption on rice husk ash". *Ind. J. Environ. Protec.*, 12,198 (Quoted by Rai, *et al.*, 1995).

26. Benefield, l. D., Judleins, J. F., Weand, B. L. (1982). "Process Chemistry for water and wastewater treatment ". *Prentice- Hall, Inc., Engle- Wood Cliffs, New Jersey.*

27. Bhalke, S., Tripathi, R. M., Mahapatra, S., Sastry, V. N., and Krishnamoorthy, T. M. (1999). "Uptake of heavy metals by sunflower plant dry powder". *Research Jour. of Chemistry and Environment*, 03, 02, 09.

28. Bharat Electronics Limited, Manufacturing processes standard. (2002). Ghaziabad, (U.P).

29. Bhattacharya, A. K. (1983). "Removal of cadmium from water by low cost adsorbents". Doctoral thesis, IIT, Kanpur, India.

30. Bird, R. B., Stewart, W. E., and Lightfoot, E. N. (1960). "Transport Phenomena". *John Wiley and Sons.*

31. Browing, E. (1969). "Toxicity of industrial metals". *Bullerworths and Co., London* (Quoted by Huang and Wu, 1975).

32. Brown, P. A., Gill, S. A., and Allen, S. J (2000). "Metal removal from wastewater using peat". *Water Research.* 34 (16), 3907-3916.

33. Calder, L. (1988). "Chromium contamination of groundwater. In: chromium in the natural and human environment". Ed: *J.O. Nriagu and E. Nieboer, Wiley series in advances in environmental science and tech.* 20:215-229.

34. Chand, S., Aggarwal, V. K., Kumar, P. (1994). " Removal of Hexavelent Chromium from the Wastewater by Adsorption". *Indian J. Environ. Hlth.* 36, 3, 151-158.

35. Cheremisinoff, P. N., and Habib, Y. H. (1972). *Wat. Sew. Wks.*, 119, 46.

36. Cimino, G., Paserini, A., and Toscano, G (2000). "Removal of toxic and Cr (VI) from aqueous solution by hazelnut shell". 34 (11), 29, 55-2962.

37. Cloutier, J. N., Leduy, A., and Ramalho, R. S. (1985). "Peat adsorption of herbicide 2, 4-D from wastewater". *Canad, J. Chem. Engg.*, 63, 250.

38. Crank, J. (1956). "The mathematics of diffusion". *Oxford Press, London.*

39. Culp, R. L., and Culp, G. L. (1971). "Advanced wastewater treatment". *Van Nostrand Reinhold Company, New York.*

40. D, Gang., Banerji, S. K., Clevenger, T. E. (1991). "Chromium (VI) removal by modified PVP- Coated Silica Gel". Proceeding of 1999 conference on hazardous waste research, 64-70.

41. De Arnab, D. (1991). "Removal of Toxic metals from water using low cost adsorbents: A review". *Journal of chemistry and environment*, 1, 2, 04, 297.

42. De Castro Dantas, T. N., Dantas Neto, A. A., and De, A. M. C. P. (2001). "Removal of chromium from aqueous solutions by diatomite treated with microemulsion" *Water Res.* 39 (9), 2219-2224.

43. De, A. K., and De, A. K. (1994). "Heavy metals removal from wastewater using flyash and agricultural wastes: A review". *Journal IAEM*, 21, 36-39.

44. Deepak, D., and Ajay, K. G. (1991). "Hexavalent chromium removal from wastewater". *Indian Jour. of Envion. Hlth.* 33, 02, 297.

45. Deo, N., Ali, M. (1992). "Use of a low cost material as an adsorbent in the removal of Cr (VI) from dilute aqueous solution". *Indian J. of Environ. Pollution*, 12, 6, 439-441.

46. Devies, M. S., and Cliffe, E. J. (1998). "Adsorption of metals from water using low cost adsorbents: a Review". *Jour. of Chemistry and Environment*, 1.2, 04, 517.

47. Drake, D. O., S, Lin, G. D Rayson., and Jackson, P. J. (1996). "Chemical modification and metal binding studies of *Datura innoxia*". *Env. Sc. and Tech.* 30 (1), 110-114.

48. Eckenfelder W., and Wasley, Jr. (1989). " Industrial water pollution control" *McGraw Hill Book Company.*

49. Faust, S. D., and Pearson, D. (1987). "Adsorption processes for water treatment" *Buttersworths, London.*

50. Fuerstenu, D. W. (1970). "Pure applied chemistry". 24, 135.

51. Gang, D., Banerji, S. K and Clevenger, T. E. (1999). "Chromium (VI) removal by modified PVP coated silica gel". Proceeding of Hazardous waste research conference.

52. Gardiner, W. G., and Munoz, F. (1971), *Chem. Eng.* 78, 57.

53. Gebehard, H., and Coleman, N. T. (1974). *Soil. Sci. Soc., Amer. Proc.*, 38,255.

54. Grover, M., Narayanswamy, S. (1982). "Removal of hexavalent chromium by adsorption on fly ash". Institution of Engineers. *J. Environ. Engg.*, 63, 36-39.

55. Gupta, G. S., Parasad, G., and Singh, V. N. (1990). *Water Res.* 24 (1), 45.

56. Gupta, G. S., Shukhla, S. P., Parasad, G., and Singh, V. N. (1992). *Environ. Technol.*, 13, 925.

57. Gupta, S. K., Srivastava, S. K., Mohan, D., and Sharma, S. (1997). "Design parameters for fixed bed reactors of activated carbon developed from fertilizer waste for the removal of some heavy metal ions". 17, 08, 517.

58. Gupta, V. K., Gupta, M., and Saurabh, S. (2001). "Process development for the removal of lead and chromium from aqueous solutions using red mud an aluminum industry waste". *Water Res.* 35 (5), 1125-1134.

59. Gupta, V. K., Gupta, M., and Saurabh, S. (2001). "Process development for the removal of lead and chromium from aqueous solutions using red mud an aluminum industry waste". *Water Res.* 35 (5), 1125-1134.

60. Gupta, V. K., Rastogi, A., and Dwivedi, M. K. (1997). "Process development for the removal of zinc and cadmium from the wastewater using slag - a blat furnace waste material". *Jour. of separation science and Technology*, 32, 17, 2883-2912.

61. Haribabau, E., Upadhya, Y. D., and Upadhyay, S. N. (1993a). "Removal of Phenols from effluent by fly ash". *Intern. Journal Env. Studies*, 43, 169 (Quoted by Rai, *et al.*, 1995).

62. Haribabau, E., Upadhya, Y. D., Upadhyay, S. N. (1993b). Proc. National Seminar on App. Chem. Engg. in the utilization of natural resources, *CSIR*.

63. Hassler, J. W. (1974). "Purification with activated Carbon". *Chemical publishing Company, New York.*

64. Hayes, R. B. (1982). "Carcinogenic effects of chromium". *Top. Environ, Health,* 5,221.

65. Heary, S. K., and Ray, J. (1987). "Processes affecting the remediation of chromium- contaminated sites". *Environmental Health Persp. NIH Publication.* 92, 24-40.

66. Heldren, R. F., and Keller, H. W. (1998). *Purdue Indust. Waste Conf.* 13, 418.

67. Huang, C. P., and Bowers, A. R. (1978). "The use of Activated Carbon for chromium (VI) removal". *Prog. Wat. Tech.* 10, 5/6, 45-64.

68. Huang, C. P., and Morehart, A. L. (1991). "Protein competition in Cu (II) Adsorption in fungal Mycelia". *Water Research*, 25, 1365-1375.

69. Huang, C. P., and Stumn, W. (1973). "Specific adsorption of cation on hydrous alumina". *J. Colloid interface Sci.*, 43,409.

70. Huang, C. P., and Wu, M. H. (1975). "Chromium removal by carbon adsorption". *Journal of Water Pollution Control Fed.*, 47, 2443-2446.

71. Huang, G. P., and Blankenship, D. W. (1984). *Water Res.*, 18, 37.

72. Jain, P. C., and Monica, J. (1996). *Engineering chemistry*.306.

73. Johnson, G. E., Kunka, L. M., and Field, H. (1965). "Use of coal and fly ash as adsorbents from secondary municipal effluents". *Ind. Eng. Chem. Des.Dev.*, 4 (3), 323.

74. Kainath, T. M. (1977). AICHE sym. *Ser. Water.*, 173, 1.

75. Katz, S. A. (1991). "The analytical biochemistry of chromium". In: Environmental health perspective. NIH Publication 92:13-16.

76. Khanna, P., and Malhotra, S. K. (1977). "Kinetics of phenol adsorption of fly ash" *Ind. J. Env. Hlth.*, 19, 3, 24.

77. Khare S. K., Panday K. K., Srivastava R. M., and Singh, V. N. (1987). " Removal of victoria blue from aqueous solution by flyash". *J. Chem. Tech. Biotechnol*, 39, 99.

78. Kim, J., and Joltek, J. (1977). "Chromium removal with activated carbon". *Prog. Water Tech.*, 9, 1,143-155.

79. Knocke, W. R., and Hemphill L. H (1981). "Mercury sorption by waste rubber" *Water Res.* 15, 275.

80. Krishnaiah, A., Nagesh, N. (1989). "A study on the removal of chromium by adsorption on activated charcoal from synthetic effluents". *Indian. J. of environ. Poll.*, 9, 4, 301-303.

81. Kumar, S. (1987). "Ph.D. theses".Banarus Hindu University, Varanasi, India.

82. Kumar, S., Upadhyay, S. N., and Upadhya, Y. D. (1987). *J. Chem. Tech. Bio.* 37,281.

83. Lal, J., Singh, D. K. (1992). "Removal of chromium (VI) from the Aqueous Solution using Waste tealeaves Carbon". *Indian J. Environ.Hlth* 34, 2, 108-113.

84. Lancy, L. E. (1967). "Philosophy and methodology of metallic waste treatment". *Plating*, 54,157 (Quoted by Patterson and Minear, 1985).

85. Landigran R. B., Hallowell, J. B. (1975). "Removal of chromium from plating rinse water using activated carbon". EPA, 670/2-75-055.

86. Lavine, S., and Smith, A. L. (1971). Discussion. *Faraday Soc.*, 52, 290.

87. Lazlo, J. A. (1966). "Preparing an ion exchange resin from sugarcane bagasse to remove reactive dye from wastewater". *Textile chemist and colorist*, 44 (5), 13-17.

88. Lee, C. K., and Low, K. S. (1989). "Removal of copper from solution using moss". *Envir. Technol. Lett*, 10,395-404.

89. Leibig, C. F. Jr., Vanselow, A. P., and Chapman, H. D. (1943). *Soil Sci.* 55, 371 (Quoted by Patterson and Minear, 1985).

90. Mahesh, S., Rama, B. M., Praveen, K. N. H., and Usha, L. K. (1999). "Adsorption kinetics of dihydric phenol hydroquinone on activated carbon". *Indian J. Environ. Hlth* 41, 4, 317-325.

91. Mahrotra, R., and Dwevedi, N. N. (1988). "Removal of Cr (VI) from water using unconventional materials". *Jour. of Indian water works association*, 20,04, 323-327.

92. Mall, I. D. (1992). "Studies on the treatment of pulp and paper mill effluent, Ph. D. theses". Banarus Hindu University, Varanasi, India (Quoted by Rai, *et al.*, 1995).

93. Mall, I.D., and Upadhyay, S. N. (1995). "Removal of basic dyes from wastewater using boiler bottom ash". *Indian J. Environ. Hlth.* 37 (1) 1-10.

94. Manju, G. N., and Anirudhan, T. S. (1990). "Use of coconut fibre pith based pseudo activated carbon for chromium removal". *Indian Jour. of Envion. Hlth*, 32, 03, 289.

95. Maranone, E., and Sastree, H. (1992). "Heavy metal removal in packed beds using apple wastes". *Biores. Technol.*, 38, 39-43.

96. Maruyama, T., Hannah, S. A., and Cohen, J. M. (1972). "Removal of heavy metals by physical and chemical treatment processes". *45th Ann. Conf., Water Poll. Control Fed.*

97. McCabe, W. R., Smith, J. U., and Harriott, P. (1993). "Unit Operations of Chemical Engineering". Fifth edition, *McGraw- Hill International Editions.*

98. Mckay, G., and Bino, M. J. (1990). "Env. Pollution". 66, 33.

99. Mckay, G., J. (1982). *Chem. Tech. Biotechnol.*, 32, 759 (Quoted by Benefield, *et al.*, 1982).

100. Mckay, G., Otterburn, M.S., and Sweeney, A.G. (1980). "The removal of colour from effluent using various adsorbents- III, Silica rate processes". *Water Res.*, 14, 15 (Quoted by Rai, *et al.*, 1995).

101. Metal-Finishing Industry Action Committee. (1953). " Method for treating metal finishing wastes," Ohio River valley water Sanitation Commission, 408 – 409.

102. Morris, J. C., and Weber, W.J. (1964). "Adsorption of Biochemically Resistant Materials from solution". *Env. Health Series AWRT-9.*

103. Naik, M. A. (2002). "Environmental Management including Waste minimization for electroplating industry". *Environmental Pollution Control Journal*, 5, 2, 41-44.

104. Nakano, Y., Tanaka, M., Nakamura, Y., and Konno, M. (2000). "Removal and recovery system of hexavalent chromium from waste water by tannin gel particles". *Journal of chem. Engg. of Japan*, 33, 5, 747-752.

105. Namasivayan, C., and Senthil, K. (1997). "Fe (III), Cr (VI) hydroxide, a waste byproduct obtained from the treatment of Cr (VI) containing wastewater in a fertilizer industry". *Jour. of Waste and Environment Today*, 16, 06, 01.

106. Navarro, R. R., Tatsumi, K., Sumi, K., and Matsumura, M. (2001). "Role of anions on heavy metal sorption of cellulose modified with poly and polyethyleneimine". *Water Research*, 35 (11), 2724-2730.

107. Neuhof, Taunusstein (1989). "Waste water technology origin collection, treatment and analysis of wastewater". Edited by Institute Fresenius GmbH, Compy Sigh by *Springer-Verlog Berlin Heilded*, 71 – 73.

108. Nieboer, E., and Jusys A. A. (1988). "Biologic chemistry of Chromium, In: chromium in the natural and human environment". Ed: J.O. Nriagu and E. Nieboer, *Wiley Series in Advances in Environmental Science and Tech.* 20:21-67.

109. Okiemen F. E., Okundia, E. V., and Ogbeifun, D.E. (1991). "Adsorption of Cd and Pb on modified groundnut husks". *J. Chem. Tech. Biotechnol.*, 51, 97-103.

110. Orhan, Y., and Byukgungor. (1993). "The removal of heavy metals by using agricultural wastes". *Wat. Sci. Tech.*, 28, 247 (Quoted by Rai, *et al.*, 1995).

111. Panday, K. K. (1984). "Ph.D. theses". B.H.U, Varanasi, India.

112. Panday, K. K., Prasad, G., and Singh, V. N. (1986). "Use of Wollastonite for the treatment of Copper II rich effluents ". *Water Air and Soil Pollution*, 27 287.

113. Panday, K. K., Prasad, G., and Singh, V. N. (1984). "Removal of Cr (VI) from aqueous solution by adsorption on flyash wallastonite". *Tech. Bio Tech*, 34A, 367-374.

114. Panday, K. K., Prasad, G., and Singh, V. N. (1985). "Copper III removal from aqueous solution by fly ash". *Water Res.* 19 (7), 869.

115. Patnaik, L. N., and Das, C. P. (1995). "Removal of hexavalent chromium by blast furnace flue dust". *Indian Jour. of Envion. Hlth,* 37, 01, 19.

116. Patterson, J. W., and Minear, R. A. (1985). " Chemical methods of heavy metals removal". *Treatment Technology, Ann. Arbow, Science Pub Inc: Ann arbor.*

117. Patterson, J. W., and Passino, J. (1987). " Treatment technology for hexavalent chromium wastewater". *Treatment Technology, Ann. Arbow, Science Pub Inc.*

118. Periasamy, K., Srinivasan, K., and Murugan P. K. (1991). "Studies on chromium (VI) removal by Activated Groundnut Husk Carbon". *Indian J. Env. Hlth,* 31, 4, 433-439.

119. Pillai, C. K. S (1981). "Coconut as a resource for materials and energy in the future". *J. of Scientific and Ind. Research,* 40, 154-165.

120. Pinner, R., and V. Crowle. (1971). *Electroplat. Met. Finisg.* 3, 13 (Quoted by Patterson and Minear, 1985).

121. Plating Room Controls for Pollution Abatement. (1951). Ohio River valley water Sanitation Commission, 412 – 417.

122. Poots, V. J. P., Healy, J. J. (1976). "Removal of acid dye from effluent using naturally occurring adsorbents Peet". *Wat.Res,* 10, 1061- 67 (Quoted by Huang and Wu, 1975).

123. Rai, A. K., Upadhyay S. N., Kumar, S., and Upadhya Y.D. (1995). " Heavy Metal Pollution and its control through a cheaper method". *Journal of IAEM,* 25, 22-51.

124. Rai, D., Sass, B. M., and Moore, D. A. (1998). "Chromium (III) hydrolysis constants and solubility of chromium (III) hydroxide". *Inorg. Chem.* 26,345-349.

125. Raji, C., and Anirudhan, T. S. (1998). "Sorptive behavior of Chromium (VI) on saw dust carbon in aqueous media". *Ecol. Env. and Cons.,* 04, 1-2, 33.

126. Ramos, R. L., Martinez, A. J., and Coronado, R. M. G. (1994). "Adsorption of Chromium (VI) from aqueous solutions on activated carbon". *Wat. Sci. Tech.,* 30, 9, 191-197.

127. Rao, M., and Bhole, A. G. (2000). "Removal of chromium using low cost adsorbents". *IAEM,* 27, 291-296.

128. Rao, M., Parwate, A. V., and Bhole A. G. (2002). "Utilization of low cost adsorbents for the removal of heavy metals from wastewater". *Env. Poll. Cont. Jour.* 5(3), 12- 23.

129. Reents, A. C., and Stromquist, D. M. (1952). "Proc. Purd. Indust. Waste conf". 7, 462.

130. Richedson, E. W., Stobbe, E. D., and Bernstein, S. (1968). *Environ. Sci. Techol,* 2, 1006 (Quoted by Patterson and Minear, 1985).

131. Rios, J. V., Bess-Oberto, L., Tiemann, K. J., and Gardea-Torresdey, J. L (1999). "Investigation of metal binding by agricultural by products". Proceeding of 1999 Conference on Hazardous Waste Research, 121-130.

132. Rothstein, S. (1958). *Plating*, 45, 835 (Quoted by Patterson and Minear, 1985).

133. Saeed, M. Y. (1999). "Management of Hazardous waste from Lock Industry in Aligarh". B. Tech project.

134. Sarvanane, R., Sundarrajan, T., and Sivammurthy, R. S. (1998). "Studies of the removal of heavy metals from wastewater using chemically modified low cost adsorbents". *Journal of the Indian Public Health engineers*, Indian, 02, 45.

135. Sarvanane, R., Sundarrajan, T., and Sivammurthy, R. S. (2002). "Efficiency of chemically modified low cost adsorbents for the removal of heavy metals from wastewater". *Indian Jour. of Envion. Hlth*, 44, 02, 78-87.

136. PSchore, G. (1972). "Electronic equipment and ion exchange for use in activated treatment systems". 27th Purdue industrial waste conference, W. Lafayette, Indiana, 2-4.

137. Schwoyer, W. L., and Luttinger, L. B. (1972). "Dewatering of metal hydroxides". Presented at the 27th purdue Industrial waste conf., W. Lafayette, Indiana, 2-4.

138. Sharma, D. C., and Forster, C.F. (1993). "Removal of Hexavalent chromium using Sphagnum mors peat". *Wat. Res*, 27, 7, 1201-08.

139. Sharma, Y. C. (1998). "Foundry material for reclamation of wastewater. Some studies". Intern. seminar on Environ. Planning and Management, Nagpur, 516-519.

140. Sharma, Y. C., Gupta, G. S., Prasad, G and Ruainwar, D. C. (1990). "Use of Wollastonite in the removal of Ni (II) from aqueous solutions". *Water, Air and Soil Pollut.*, 49, 69.

141. Sharma, Y. C., Kaul, S. N., and Rupainwar, D.C. (1999). "Cr (VI) removal from industrial effluents by an Economic method". Pro. of conference at Nagpur, 11-26.

142. Sharma, Y. C., Prasad, G and Ruainwar, D. C. (1992). *In. J, Env. Studies*, 37, 183.

143. Sharma, Y. C., Prasad, G., and Ruainwar, D. C. (1990a). "Inexpensive adsorption technique to remove Cr (VI) from aqueous solutions". *In. J. Anal. Chem.*, 39, 301.

144. Sharma, Y.C., Prasad, G., Rupainwar, D.C. (1991). *Intern. J. Env. Studies*, 37, 183.

145. Shen, K., and Wang, J. (1995). " Hexavalent chromium removal in two stage bioreactor system". *J. Envir. Eng., ASCE*. 121 (11). 798-804.

146. Shukhla, S. R., Sakhardane, V. D. (1991). "Column studies on metal iron removal by dyed cellulosic materials". 11 (4), 284-289.

147. Shupack, S. I. (1991). "The Chemistry of chromium and some resulting analytical problems". *In: Environmental Health Persp., NIH Publication*. 92,17-24.

148. Siddiqui, Z. M., Paroor, S. (1994). "Removal of chromium (VI) by different Adsorbents –A comparative Study". *IJEP*, 14, 4, 273-278.

149. Singh A. K., Singh, D. P., Panday, K. K and Singh, V. N (1988). "Wollastonite as adsorbent for the removal of Fe (II) from water". *J. chem. Tech. Biotechnol.*, 42, 39 (Quoted by Rai, *et al.*, 1995).

150. Singh D. K., and Srivastava, B. (1999). "Removal of basic dyes from aqueous solutions by chemically treated *Psidium guajava* leaves". *Indian J. Environ Hlth* 41 (4) 333-345.

151. Singh, D. K., and Mishra, A. (1990). *Indian Jour. of Envion. Hlth.*, 32, 04, 345.

152. Singh, D. K., and Mishra, N. K. (1992). "Removal of toxic metal ions using modified adsorbent". *Journal of Pollution Research, India*, 11, 04, 187.

153. Singh, D. K., Saksena, D. N., and Tiwari, D. P. (1994). "Removal of Cr (VI) from aqueous solutions using low cost adsorbents". *Indian J. Envion. Hlth*, 36, 4, 272-277.

154. Singh, D. K., Srivastava, B., and Bharadwaj, R. K. (2001). "Removal of chromium (VI), Iron (III) and Mercury (II) for the aqueous solutions using activated carbon obtained from used tea leaves". *Pollution Res.* 20,2, 173-177.

155. Singh, R. P., Vaishya, R. C., Prasad, S. C. (1992). "A Study of removal of hexavalent chromium by adsorption on saw dust". *Ind. J. Environ. Protec.*, 12, 12, 916.

156. Singh, V. K., and Tiwari, P. N. (1997). "Removal and recovery of Chromium (VI) from industrial wastewater". *J. Chem. Tech. Biotechnl.* 69, 376-382.

157. Snoeyink, V. L., and Weber, W. J. Jr. (1967). *Env. Sci. Technol.*, 1, 3, 228.

158. Srinivasan, K., Balasubramanian, N., and Ramakrishna, T. V. (1988). "Studies on the chromium removal by rice husk carbon". *Indian J. Envion. Hlth.*, 30, 04,376.

159. Srivastava, S. K., Gupta, V. K., and Mohan, D. (1996). "Kinetic parameters for the removal of lead and chromium from wastewater using activated carbon developed fertilizer waste material". *j. Environmental Modeling and Assessment*, 01, 281.

160. Standard Methods for the analysis water and wastewater. (1989). 17th ed. APH, AWWA, WPCF, Washington D.C.

161. Stanley, E. (1993). Environmental Chemistry, Manhan Compy Sigh © *Lewish Publishers*, 285 – 286.

162. Stomberg, A. (1984). *Argon. Journal*, 76,719 (Quoted by Patterson and Minear, 1985).

163. Stone, E. H. F. (1967). Proc. Purdue Ind. Waste Conf., 22, 848 (Quoted by Patterson and Minear, 1985).

164. Stummm W., Morgon J. J. (1970). Aquatic chemistry, *John Wiley, N. Y.*

165. Stumn, W., and Bilinski, H. (1972). Proc. 6th Int. Conf. On water pollut. Res., pergamon press, New York.

166. Sujata, K. M., Surendra, K., and Upadhyay, S. N. (1995). "Chromium (VI) removal by spent myrobalan nuts". *J. of Indian Association for Env. Management*, 22, 247.

167. Tare, V., Jawed, M., and Leela, L. (1988). "Adsorption of Xanthates in heavy metal removal". *Journal of Indian Association for Water Pollution Tech.* Annual, 15, 88-94.

168. Tiwari, D. P., Promad, I. C., Mishra, A. K., Singh, R. P., and Srivastava, R. P. S. (1989). "Removal of toxic metals from electroplating industries (Effect of pH on removal by adsorption)". *Indian Jour. of Envion. Hlth*, 31, 02,120-124.

169. Tondon, S. K. (1982). *Top. Env. Hlth*, 5,209 (Quoted by Patterson and Minear, 1985).

170. Tran, H.H., Roddick, F. A., and O' Donnel, J. A. (1999). "Comparison of chromatography and desiccant silica gels for the adsorption of metal ions -I. Adsorption and kinetics". *Water Research* 33 (13), 2992-3000.

171. Vaishya, R., and Prasad, S. C. (1991). "Adsorption of copper on saw dust". *Indian J. Envoron. Prot.*, 11, 284-299.

172. Vanagamudi, A., Kannan. (1991). "A study on removal of Cr (VI) by adsorption on lignite coal". *Indian jour. of Environt. Pollution*, 11, 4, 241-45.

173. Vasanth, K. K., and Subanandam, K. (2002). "Studies on decolorisation of basic dye on to carbonized waste adsorbent". *Indian jour. of Environ. Pollution*, 5, 2, 5-11.

174. Waldsex, V. R., and Jaycock, P. (1971). "Removals of heavy metals". *J. Envir. Technol. Lett*, 12, 45-51.

175. Waste water technology origin collection. (1989). "Treatment and analysis of wastewater" Edited by Institute Fresenius GmbH, Taunusstein-Neuhof. Compy Sigh by *Springer-Verlog Berlin Heilded*, 71 – 73.

176. Weber, C.W. (1996). "In Vitro binding capacity of wheat bran, rice bran and oat fiber for Ca, Mg, Cu and Zn alone and in different combinations". *Journal of Agric. Food Chem*, 44, 2067-2072.

177. Weber, W. J. Jr. (1972). "Physico - Chemical processes for water quality control". *Wiley Inter Science*, New York.

178. Weber, W. J. Jr., and Moris, J. C. (1963a). "Kinetics of adsorption on carbon from solution". *J. Sant. Eng. Div., ASCE*, 890, Sa2, 31.

179. Weber, W. J. Jr., and Moris, J. C. (1963b). "Kinetics of adsorption on carbon from solution". *J. Sant. Eng. Div., ASCE*, 89, SA6, 53.

180. Weber, W. J. Jr., and Moris, J. C. (1964). *J. Sant. Eng. Div., Proc. Amer. Soc., Civil Engr.*, 90, SA3, 79 (Quoted by Benefield, *et al.*, 1982).

181. Wise, W. B.F. Dodge., and H. Bliss. (1947). "Brass and Copper Industry", Industrial and Engineering Chemistry, 407.

182. Wright, E. M., and Pratt, N. C. (1974). *J. Chem. Soc., Faraday Transaction*, 70, 1, 1461 (Quoted by Benefield, *et al.*, 1982).

183. Wright, H. J. L., and Heunter, R. J. (1973). *Aust. J. Chem.*, 26, 1183 (Quoted by Patterson and Minear, 1985).

184. Yoshio, N., Takeshita, K., and Tsutsumi, T. (2001). "Adsorption mechanism of hexavalent chromium by redox within condensed -tannin gel". *W. Research*, 35, 2, 496-500.

185. Yuronis, D. (1968). *Journal of Plating*, 55, 1071 (Quoted by Patterson and Minear, 1985).

Appendices

Appendix A

Data for Coconut Shell

(a) Effect of Adsorbent Dose and Contact Time on Fraction Adsorbed

Contact time	**1 hr**
Adsorbent	Coconut shell
Initial chromium conc.(Ci)	50mg/l
pH	1.5
Size of adsorbent	150 micron
Temperature	30° C

Adsorbent Dose (g)	*Conc. of Cr(VI) after Adsorption, mg/l (Ci)*	*Conc. of Cr(VI) Adsorbed, mg/l (Ci-Ce)*	*Friction Adsorbed [(Ci-Ce)/Ci]*
2.0	41.5	8.5	0.17
6.0	32.5	17.5	0.35
10	24.35	26.6	0.55
20	16	34	0.68
30	14	36	0.72
40	12.5	37.5	0.75
50	11.5	38.5	0.77
60	10.10	39.9	0.79

Contact time	**5.0 hr**
Adsorbent	Coconut shell
Initial chromium conc.(Ci)	50mg/l
pH	1.5
Size of adsorbent	150 micron
Temperature	30° C

Adsorbent Dose (g)	*Conc. of Cr(VI) after Adsorption, mg/l (Ce)*	*Conc. of Cr(VI) Adsorbed, mg/l (Ci-Ce)*	*Friction Adsorbed [(Ci-Ce)/Ci]*
2.0	37.5	12.5	0.25
6.0	29.5	20.5	0.41
10	16	34	0.68
20	12.5	37.5	0.75
30	10	40	0.80
40	9.5	40.5	0.81
50	6.6	43.4	0.87
60	5.8	44.2	0.88

Contact time	**12 hr**
Adsorbent	Coconut shell
Initial chromium conc.(Ci)	50mg/l
pH	1.5
Size of adsorbent	150 micron
Temperature	30° C

Adsorbent Dose (g)	*Conc. of Cr(VI) after Adsorption, mg/l (Ce)*	*Conc. of Cr(VI) Adsorbed, mg/l (Ci-Ce)*	*Friction Adsorbed [(Ci-Ce)/Ci]*
2.0	36	14	0.28
6.0	21	29	0.58
10	12.5	37.5	0.75
20	9.5	40.5	0.81
30	5.0	45.0	0.907
40	2.6	47.4	0.948
50	1.75	48.25	0.965
60	1.75	48.25	0.965

Contact time	**24 hr**
Adsorbent	Coconut shell
Initial chromium conc.(Ci)	50mg/l
pH	1.5
Size of adsorbent	150 micron
Temperature	30° C

Adsorbent Dose (g)	*Conc. of Cr(VI) after Adsorption, mg/l (Ce)*	*Conc. of Cr(VI) Adsorbed, mg/l (Ci-Ce)*	*Friction Adsorbed [(Ci-Ce)/Ci]*
2.0	32.0	18.0	0.36
6.0	19.0	31.0	0.62
10	7.95	42.05	0.841
20	2.50	47.50	0.95
30	0.55	49.45	0.989
40	0.30	49.70	0.994
50	0	50.0	1.00
60	0	50.0	1.00

Influence of pH on Fraction Adsorbed

Adsorbent	**Coconut shell**
Adsorbent dose	10.0g/l
Contact time	24 hr
Initial Cr Conc.(Ci)	50mg/l
Temperature	30° C
Size of adsorbent	150 micron

pH of the Solution	*Conc. of Cr(VI) after Adsorption, mg/l (Ce)*	*Conc. of Cr(VI) Adsorbed, mg/l (Ci-Ce)*	*Friction Adsorbed [(Ci-Ce)/Ci]*
1	7.5	42.5	0.85
2.2	16.0	34.0	0.76
3	24.0	26.0	0.52
6	31.5	18.5	0.37
7	34.0	16.0	0.32
9.2	35.0	15.0	0.3
10	35.0	15.0	0.3

Effect of Initial Cr (VI) Concentration on Fraction Adsorbed

Adsorbent	**Coconut shell**
Contact time	24 hr
Adsorbent dose	10.0 g/l
Temperature	30° C
Size of adsorbent	150 micron
pH	1.5

Initial Cr(VI) Conc. before Adsorption, (mg/l)	*Conc. of Cr(VI) after Adsorption, mg/l (Ce)*	*Conc. of Cr(VI) Adsorbed, mg/l (Ci-Ce)*	*Friction Adsorbed [(Ci-Ce)/Ci]*
5	0	5.0	1.0
10	0.5	9.5	0.95
20	2.1	17.9	0.90
30	3.2	26.8	0.89
40	4.8	35.2	0.88
50	9.5	40.5	0.81
60	17.4	42.5	0.72
70	24.0	46.0	0.65
80	32.0	48.0	0.60
90	42.0	48.0	0.53
100	50.0	50.0	0.50

Effect of Contact Time and Particle Size on Fraction Adsorbed

Size of adsorbent	**300 micron**
Adsorbent dose	10.0 g/l
Adsorbent	Coconut shell
Initial Cr Conc.(Ci)	50mg/l
Temperature	30° C
pH	1.5

Contact Time (hr)	*Conc. of Cr(VI) after Adsorption, mg/l (Ce)*	*Conc. of Cr(VI) Adsorbed, mg/l (Ci-Ce)*	*Friction Adsorbed [(Ci-Ce)/Ci]*
0.5	37.5	12.50	0.25
1.0	32.95	17.05	0.34

Contact Time (hr)	*Conc. of Cr(VI) after Adsorption, mg/l (Ce)*	*Conc. of Cr(VI) Adsorbed, mg/l (Ci-Ce)*	*Friction Adsorbed [(Ci-Ce)/Ci]*
5.0	22.9	27.1	0.542
9.0	19.0	31.0	0.62
12.0	16.75	33.25	0.66
24.0	13.79	36.2	0.724

Size of adsorbent	**150 micron**
Adsorbent dose	10.0 g/l
Adsorbent	Coconut shell
Initial Cr Conc.(Ci)	50mg/l
Temperature	30° C
pH	1.5

Contact Time (hr)	*Conc. of Cr(VI) after Adsorption, mg/l (Ce)*	*Conc. of Cr(VI) Adsorbed, mg/l (Ci-Ce)*	*Friction Adsorbed [(Ci-Ce)/Ci]*
0.5	33.31	16.69	0.333
1.0	28.95	21.05	0.421
5.0	23.25	26.75	0.535
9.0	14.95	35.05	0.701
12.0	10.74	39.26	0.785
24.0	7.15	42.85	0.857

Size of adsorbent	**75 micron**
Adsorbent dose	10.0 g/l
Adsorbent	Coconut shell
Initial Cr Conc.(Ci)	50mg/l
Temperature	30° C
pH	1.5

Contact Time (hr)	Conc. of Cr(VI) after Adsorption, mg/l (Ce)	Conc. of Cr(VI) Adsorbed, mg/l (Ci-Ce)	Friction Adsorbed [(Ci-Ce)/Ci]
0.5	28.85	21.15	0.423
1.0	24.0	26.0	0.52
5.0	14.45	35.55	0.711
9.0	7.90	42.10	0.842
12.0	6.65	43.35	0.867
24.0	5.55	44.45	0.889

Effect of Contact Time and Temperature on Fraction Adsorbed

Temperature	**30° C**
Adsorbent dose	10.0 g/l
Adsorbent	Coconut shell
Initial Cr Conc.(Ci)	50mg/l
Size of adsorbent	150 micron
pH	1.5

Contact Time (hr)	Conc. of Cr(VI) after Adsorption, mg/l (Ce)	Amount of Solute Adsorbed mg(x)	Amount Adsorbed Qe = x/m (mg/g)	Friction Adsorbed [(Ci-Ce)/Ci]
0.5	33.31	16.69	1.669	33.38
2.0	30.80	19.19	1.91	38.40
3.0	28.2	21.81	2.18	43.60
5.0	23.45	26.55	2.65	53.10
9.0	15.38	34.62	3.46	69.24
12.0	10.74	39.26	3.92	78.52
15.0	8.12	41.88	4.18	83.76
24.0	7.121	42.87	4.28	85.75

Temperature	**40° C**
Adsorbent dose	10.0 g/l
Adsorbent	Coconut shell
Initial Cr Conc.(Ci)	50mg/l
Size of adsorbent	150 micron
pH	1.5

Contact Time (hr)	*Conc. of Cr(VI) after Adsorption, mg/l (Ce)*	*Amount of Solute Adsorbed mg(x)*	*Amount Adsorbed Qe = x/m (mg/g)*	*Friction Adsorbed [(Ci-Ce)/Ci]*
0.5	28.46	21.54	2.154	43.08
2.0	26.16	23.84	2.38	47.68
3.0	23.98	26.02	2.60	52.04
5.0	19.04	30.96	3.09	61.92
9.0	11.90	38.10	3.81	76.20
12.0	8.14	41.86	4.18	83.72
15.0	6.10	43.90	4.39	87.80
24.0	5.88	44.12	4.41	88.24

Temperature	**50° C**
Adsorbent dose	10.0 g/l
Adsorbent	Coconut shell
Initial Cr Conc.(Ci)	50mg/l
Size of adsorbent	150 micron
pH	1.5

Contact Time (hr)	*Conc. of Cr(VI) after Adsorption, mg/l (Ce)*	*Amount of Solute Adsorbed mg(x)*	*Amount Adsorbed Qe = x/m (mg/g)*	*Friction Adsorbed [(Ci-Ce)/Ci]*
0.5	24.18	25.82	2.58	51.64
2.0	21.54	28.46	2.84	56.92
3.0	19.48	30.52	3.05	61.04
5.0	14.96	35.04	3.50	70.08
9.0	8.10	41.90	4.19	83.80
12.0	5.11	44.89	4.48	89.78
15.0	4.12	45.88	4.58	91.76
24.0	3.89	46.11	4.61	92.22

Freundlich Adsorption Isotherms

Temperature **30° C**

Weight of the Adsorbent, mg (m)	*Conc. of Cr(VI) after Adsorption, mg/l (Ce)*	*Amount of Solute Adsorbed, mg(x)*	*Amount Adsorbed qe = x/m (mg/mg)*	*Log Ce*	*Log qe*
2000	33.31	16.69	8.34×10^{-3}	1.522	–2.078
6000	19.321	30.679	5.113×10^{-3}	1.286	–2.495
10000	7.121	42.879	4.287×10^{-3}	0.852	–2.367
20000	5.00	45.0	2.25×10^{-3}	0.698	–2.70

Temperature **40° C**

Weight of the Adsorbent, mg (m)	*Conc. of Cr(VI) after Adsorption, mg/l (Ce)*	*Amount of Solute Adsorbed, mg(x)*	*Amount Adsorbed qe = x/m (mg/mg)*	*Log Ce*	*Log qe*
2000	31.957	18.043	9.02×10^{-3}	1.504	–2.088
6000	10.230	39.77	6.627×10^{-3}	1.009	–2.179
10000	8.84	41.16	4.11×10^{-3}	0.946	–2.40
20000	4.952	45.048	2.24×10^{-3}	0.694	–2.649

Temperature **50° C**

Weight of the Adsorbent, mg (m)	*Conc. of Cr(VI) after Adsorption, mg/l (Ce)*	*Amount of Solute Adsorbed, mg(x)*	*Amount Adsorbed qe = x/m (mg/mg)*	*Log Ce*	*Log qe*
2000	31.640	18.36	9.175×10^{-3}	1.500	–2.017
6000	9.645	40.335	6.725×10^{-3}	0.984	–2.172
10000	4.651	45.349	4.53×10^{-3}	0.667	–2.343
20000	3.33	46.67	2.33×10^{-3}	0.522	–2.632

Langmuir Adsorption Isotherms

Temperature **30° C**

Weight of the Adsorbent, mg (m)	*Conc. of Cr(VI) after Adsorption, mg/l (Ce)*	*Amount of Solute Adsorbed, mg(x)*	*Amount Adsorbed qe = x/m (mg/mg)*	$\frac{1}{x/m}$	$\frac{1}{Ce}$
2000	33.31	16.69	8.34 x10^{-3}	119.90	0.030
6000	19.321	30.679	5.113 x10^{-3}	195.694	0.0517
10000	7.121	42.879	4.287 x10^{-3}	233.263	0.140
20000	5.00	45.0	2.25x10^{-3}	444.45	0.200

Temperature **40° C**

Weight of the Adsorbent, mg (m)	*Conc. of Cr(VI) after Adsorption, mg/l (Ce)*	*Amount of Solute Adsorbed, mg(x)*	*Amount Adsorbed qe = x/m (mg/mg)*	$\frac{1}{x/m}$	$\frac{1}{Ce}$
2000	31.957	18.043	9.02 x10^{-3}	110.85	0.031
6000	10.230	39.77	6.627 x10^{-3}	150.89	0.097
10000	8.84	41.16	4.11 x10^{-3}	243	0.113
20000	4.952	45.048	2.24 x10^{-3}	355	0.195

Temperature **50° C**

Weight of the Adsorbent, mg (m)	*Conc. of Cr(VI) after Adsorption, mg/l (Ce)*	*Amount of Solute Adsorbed, mg(x)*	*Amount Adsorbed qe = x/m (mg/mg)*	$\frac{1}{x/m}$	$\frac{1}{Ce}$
2000	31.640	18.36	9.175 x10^{-3}	108.98	0.031
6000	9.645	40.335	6.725 x10^{-3}	132.84	0.103
10000	4.651	45.349	4.53 x10^{-3}	198.53	0.201
20000	3.33	46.67	2.33 x10^{-3}	297	0.300

Kinetics of Adsorption

Adsorbent dose	10.0 g/l
Adsorbent	Coconut shell
Initial Cr Conc.(Ci)	50mg/l
Temperature	30° C

Contact Time (hr)	*Conc. of Cr(VI) after Adsorption, mg/l (Ce)*	*Amount of Solute Adsorbed mg(x)*	*Amount Adsorbed Qe = x/m (mg/g)*	*Log (ge - q)*
0.5	33.31	16.69	1.669	0.416
2.0	30.80	19.19	1.91	0.373
3.0	28.2	21.81	2.18	0.322
5.0	23.45	26.55	2.65	0.210
9.0	15.38	34.62	3.46	–0.0872
12.0	10.74	39.26	3.92	–0.4509
15.0	8.12	41.88	4.18	–1.036
24.0	7.121	42.87	4.28	–

Thermodynamic Parameters

Temperature in Kelvin	*Langmuir Constant*	*Log b*	$\frac{1}{T}$
303	0.0220	–1.657	3.3 0 x10[3]
313	0.0244	–1.612	3.19 x10[3]
323	0.0315	–1.501	3.09 x10[3]

Inter Particle Diffusion

Adsorbent dose	10.0 g/l
Adsorbent	Coconut shell
Initial Cr Conc.(Ci)	50mg/l
Temperature	30° C
Size of adsorbent	150 micron

Contact Time (mint)	*Conc. of Cr(VI) after Adsorption, mg/l (Ce)*	*Amount of Solute Adsorbed mg(x)*	*Amount Adsorbed Qe = x/m (mg/g)*	$\sqrt{t}$ *(mint)*
30	33.31	16.69	1.669	5.42
120	30.80	19.19	1.91	10.95
180	28.2	21.81	2.18	13.41
300	23.45	26.55	2.65	17.32
540	15.38	34.62	3.46	23.23
600	10.74	39.26	3.92	24.49
900	8.12	41.88	4.18	30.00
1440	7.121	42.87	4.28	37.94

Appendix B

Data for Neem Bark

Effect of Adsorbent Dose and Contact Time on Fraction Adsorbed

Contact time	**1 hr**
Adsorbent	Neem Bark
Initial chromium conc.(Ci)	50mg/1
pH	3
Size of adsorbent	150 micron
Temperature	30° C

Adsorbent Dose (g)	*Conc. of Cr(VI) after Adsorption, mg/l (Ci)*	*Conc. of Cr(VI) Adsorbed, mg/l (Ci-Ce)*	*Friction Adsorbed [(Ci-Ce)/Ci]*
2	43.8	6.20	0.124
4	37.6	12.4	0.248
6	32.0	18.0	0.360
8	27.6	22.4	0.448
10	23.45	26.55	0.531
12	21.05	28.95	0.579
14	18.6	31.4	0.628
16	16.9	33.1	0.662
18	15.5	34.5	0.69
20	14.5	35.5	0.71

Contact time	**6.0 hr**
Adsorbent	Neem Bark
Initial chromium conc.(Ci)	50mg/1
pH	3
Size of adsorbent	150 micron
Temperature	30° C

Adsorbent Dose (g)	*Conc. of Cr(VI) after Adsorption, mg/l (Ce)*	*Conc. of Cr(VI) Adsorbed, mg/l (Ci-Ce)*	*Friction Adsorbed [(Ci-Ce)/Ci]*
2	39.95	10.05	0.201
4	31.9	18.10	0.363
6	25.45	24.55	0.491
8	21.05	28.95	0.579
10	17.25	32.75	0.655
12	13.95	36.05	0.721
14	12.40	37.60	0.752
16	11.05	38.95	0.779
18	10.7	39.30	0.786
20	10.7	39.30	0.786

Contact time	**12 hr**
Adsorbent	Neem Bark
Initial chromium conc.(Ci)	50mg/1
pH	3
Size of adsorbent	150 micron
Temperature	30° C

Adsorbent Dose (g)	*Conc. of Cr(VI) after Adsorption, mg/l (Ce)*	*Conc. of Cr(VI) Adsorbed, mg/l (Ci-Ce)*	*Friction Adsorbed [(Ci-Ce)/Ci]*
2	34.25	15.75	0.315
4	24.85	25.15	0.503
6	17.95	32.05	0.641
8	13.10	36.9	0.738
10	9.95	40.05	0.801

Adsorbent Dose (g)	*Conc. of Cr(VI) after Adsorption, mg/l (Ce)*	*Conc. of Cr(VI) Adsorbed, mg/l (Ci-Ce)*	*Friction Adsorbed [(Ci-Ce)/Ci]*
12	9.05	40.95	0.819
14	8.45	41.55	0.813
16	7.90	42.10	0.842
18	7.20	42.8	0.856
20	7.20	42.8	0.856

Influence of pH on Fraction Adsorbed

Adsorbent	**Neem Bark**
Adsorbent dose	10.0g/l
Contact time	24 hr
Initial Cr Conc.(Ci)	50mg/l
Temperature	30° C
Size of adsorbent	150 micron

pH of the Solution	*Conc. of Cr(VI) after Adsorption, mg/l (Ce)*	*Conc. of Cr(VI) Adsorbed, mg/l (Ci-Ce)*	*Friction Adsorbed [(Ci-Ce)/Ci]*
1	32.90	17.10	0.342
2	22.0	28.0	0.56
3	9.75	40.25	0.806
4	10.75	39.25	0.785
7	12.45	37.55	0.751
9	17.10	32.9	0.658

Effect of Initial Cr (VI) Concentration on Fraction adsorbed

Adsorbent	**Neem Bark**
Contact time	12 hr
Adsorbent dose	10.0 g/l
Temperature	30° C
Size of adsorbent	150 micron
pH	3

Initial Cr(VI) Conc. before Adsorption, (mg/l)	*Conc. of Cr(VI) after Adsorption, mg/l (Ce)*	*Conc. of Cr(VI) Adsorbed, mg/l (Ci-Ce)*	*Friction Adsorbed [(Ci-Ce)/Ci]*
5	0.0	50.0	1.0
10	0.9	49.1	0.982
20	3.45	46.55	0.931
30	5.45	44.55	0.891
40	6.85	43.15	0.863
50	9.80	40.20	0.804
60	13.6	36.4	0.728
70	17.15	32.85	0.657
80	20.40	29.6	0.592
100	25.10	24.90	0.498

Effect of Contact Time and Particle Size on Fraction Adsorbed

Size of adsorbent	**300 micron**
Adsorbent dose	10.0 g/1
Adsorbent	Neem Bark
Initial Cr Conc.(Ci)	50mg/1
Temperature	30° C
pH	3

Contact Time (hr)	*Conc. of Cr(VI) after Adsorption, mg/l (Ce)*	*Conc. of Cr(VI) Adsorbed, mg/l (Ci-Ce)*	*Friction Adsorbed [(Ci-Ce)/Ci]*
1	37.95	12.05	0.241
3	20.40	29.6	0.592
6	15.45	34.55	0.691
12	13.90	36.10	0.722
18	12.40	37.60	0.752

Size of adsorbent	**150 micron**
Adsorbent dose	10.0 g/l
Adsorbent	Neem Bark
Initial Cr Conc.(Ci)	50mg/l
Temperature	30° C
pH	3

Contact Time (hr)	*Conc. of Cr(VI) after Adsorption, mg/l (Ce)*	*Conc. of Cr(VI) Adsorbed, mg/l (Ci-Ce)*	*Friction Adsorbed [(Ci-Ce)/Ci]*
1	34.4	15.6	0.312
3	17.15	32.8	.657
6	10.8	39.2	0.784
12	9.45	40.55	0.811
18	8.75	41.25	0.825

Size of adsorbent	**75 micron**
Adsorbent dose	10.0 g/l
Adsorbent	Neem Bark
Initial Cr Conc.(Ci)	50mg/l
Temperature	30° C
pH	3

Contact Time (hr)	*Conc. of Cr(VI) after Adsorption, mg/l (Ce)*	*Conc. of Cr(VI) Adsorbed, mg/l (Ci-Ce)*	*Friction Adsorbed [(Ci-Ce)/Ci]*
1	27.2	22.8	0.456
3	13.1	36.9	0.738
6	7.15	42.85	0.857
12	4.65	45.35	0.907
18	4.25	45.75	0.915

Effect of Contact Time and Temperature on Fraction Adsorbed

Temperature	**30° C**
Adsorbent dose	10.0 g/l
Adsorbent	Neem Bark
Initial Cr Conc.(Ci)	50mg/l
Size of adsorbent	150 micron
pH	3

Contact Time (hr)	*Conc. of Cr(VI) after Adsorption, mg/l (Ce)*	*Amount of Solute Adsorbed mg(x)*	*Amount Adsorbed Qe = x/m (mg/g)*	*Friction Adsorbed [(Ci-Ce)/Ci]*
1	34.4	15.6	1.56	0.312
3	17.15	32.8	3.285	0.657
6	10.8	39.2	3.92	0.784
12	9.45	40.55	4.055	0.811
18	8.75	41.25	4.125	0.825

Temperature	**40° C**
Adsorbent dose	10.0 g/l
Adsorbent	Neem Bark
Initial Cr Conc.(Ci)	50mg/l
Size of adsorbent	150 micron
pH	3

Contact Time (hr)	*Conc. of Cr(VI) after Adsorption, mg/l (Ce)*	*Amount of Solute Adsorbed mg(x)*	*Amount Adsorbed Qe = x/m (mg/g)*	*Friction Adsorbed [(Ci-Ce)/Ci]*
1	28.6	21.4	2.14	0.428
3	11.8	38.2	3.82	0.765
6	7.86	42.14	4.214	0.843
12	6.70	43.30	4.330	0.866
18	6.66	43.34	4.334	0.867

Temperature	**50° C**
Adsorbent dose	10.0 g/l
Adsorbent	Neem Bark
Initial Cr Conc.(Ci)	50mg/l
Size of adsorbent	150 micron
pH	3

Contact Time (hr)	*Conc. of Cr(VI) after Adsorption, mg/l (Ce)*	*Amount of Solute Adsorbed mg(x)*	*Amount Adsorbed Qe = x/m (mg/g)*	*Friction Adsorbed [(Ci-Ce)/Ci]*
1	23.38	26.62	2.662	0.5324
3	9.765	40.235	4.023	0.8047
6	6.66	43.34	4.334	0.8668
12	4.93	45.07	4.507	0.9014
18	4.92	45.08	4.508	0.9016

Freundlich Adsorption Isotherms

Temperature **30° C**

Weight of the Adsorbent, mg (m)	*Conc. of Cr(VI) after Adsorption, mg/l (Ce)*	*Amount of Solute Adsorbed, mg(x)*	*Amount Adsorbed qe = x/m (mg/mg)*	*Log Ce*	*Log qe*
2000	34.25	15.75	7.875×10^{-3}	1.534	–2.103
4000	24.85	25.15	6.287×10^{-3}	1.395	–2.201
6000	17.95	32.05	5.341×10^{-3}	1.254	–2.27
8000	13.10	36.90	4.612×10^{-3}	1.117	–2.33
10000	9.95	40.05	4.005×10^{-3}	0.997	–2.397

Temperature **40° C**

Weight of the Adsorbent, mg (m)	*Conc. of Cr(VI) after Adsorption, mg/l (Ce)*	*Amount of Solute Adsorbed, mg(x)*	*Amount Adsorbed qe = x/m (mg/mg)*	*Log Ce*	*Log qe*
2000	30.544	19.456	9.728×10^{-3}	1.484	–2.011
4000	20.824	29.176	7.294×10^{-3}	1.318	–2.137
6000	11.888	38.112	6.352×10^{-3}	1.075	–2.197
8000	7.92	42.08	5.26×10^{-3}	0.898	–2.279
10000	7.121	42.879	4.288×10^{-3}	0.852	–2.367

Temperature **50° C**

Weight of the Adsorbent, mg (m)	*Conc. of Cr(VI) after Adsorption, mg/l (Ce)*	*Amount of Solute Adsorbed, mg(x)*	*Amount Adsorbed qe = x/m (mg/mg)*	*Log Ce*	*Log qe*
2000	25.732	24.268	12.31×10^{-3}	1.410	–1.91
4000	12.496	37.504	9.376×10^{-3}	1.096	–2.027
6000	4.70	45.300	7.55×10^{-3}	0.672	–2.12
8000	3.44	46.56	5.82×10^{-3}	0.536	–2.35
10000	3.402	46.598	4.669×10^{-3}	0.531	–2.331

Langmuir Adsorption Isotherms

Temperature **30° C**

Weight of the Adsorbent, mg (m)	*Conc. of Cr(VI) after Adsorption, mg/l (Ce)*	*Amount of Solute Adsorbed, mg(x)*	*Amount Adsorbed qe = x/m (mg/mg)*	$\frac{1}{x/m}$	$\frac{1}{Ce}$
2000	34.25	15.75	7.875×10^{-3}	127.06	0.0291
4000	24.85	25.15	6.287×10^{-3}	159.23	0.0402
6000	17.95	32.05	5.341×10^{-3}	187.26	0.0557
8000	13.10	36.90	4.612×10^{-3}	216.91	0.0762
10000	9.95	40.05	4.005×10^{-3}	250.0	0.1005

Temperature **40° C**

Weight of the Adsorbent, mg (m)	*Conc. of Cr(VI) after Adsorption, mg/l (Ce)*	*Amount of Solute Adsorbed, mg(x)*	*Amount Adsorbed qe = x/m (mg/mg)*	$\frac{1}{x/m}$	$\frac{1}{Ce}$
2000	30.544	19.456	9.728×10^{-3}	102.88	0.0327
4000	20.824	29.176	7.294×10^{-3}	137.17	0.0480
6000	11.888	38.112	6.352×10^{-3}	157.48	0.841
8000	7.92	42.08	5.26×10^{-3}	190.11	0.1262
10000	7.121	42.879	4.288×10^{-3}	233.64	0.1404

Temperature **50° C**

Weight of the Adsorbent, mg (m)	*Conc. of Cr(VI) after Adsorption, mg/l (Ce)*	*Amount of Solute Adsorbed, mg(x)*	*Amount Adsorbed qe = x/m (mg/mg)*	$\frac{1}{x/m}$	$\frac{1}{Ce}$
2000	25.732	24.268	12.31×10^{-3}	82.64	0.0388
4000	12.496	37.504	9.376×10^{-3}	106.72	0.0800
6000	4.70	45.300	7.55×10^{-3}	132.45	0.2127
8000	3.44	46.56	5.82×10^{-3}	171.82	0.2906
10000	3.402	46.598	4.669×10^{-3}	215.05	0.2939

Kinetics of Adsorption

Adsorbent dose **10.0 g/l**

Adsorbent Neem Bark

Initial Cr Conc.(Ci) 50mg/1

Temperature 30° C

Contact Time (hr)	*Conc. of Cr(VI) after Adsorption, mg/l (Ce)*	*Amount of Solute Adsorbed mg(x)*	*Amount Adsorbed Qe = x/m (mg/g)*	*Log (ge - q)*
1	34.4	15.6	1.56	0.409
3	17.15	32.8	3.285	–0.0731
6	10.8	39.2	3.920	–0.688
12	9.45	40.55	4.055	–1.124
18	8.75	41.25	4.125	–

Thermodynamic Parameters

Temperature in Kelvin	*Langmuir Constant*	*Log b*	$\frac{1}{T}$
303	0.0160	–1.795	3.3 0 x10^{3}
313	0.0195	–1.7099	3.19 x10^{3}
323	0.0263	–1.580	3.09 x10^{3}

Inter Particle Diffusion

Adsorbent dose	**10.0 g/l**
Adsorbent	Neem Bark
Initial Cr Conc.(Ci)	50mg/1
Temperature	30° C
Size of adsorbent	150 micron

Contact Time (mint)	*Conc. of Cr(VI) after Adsorption, mg/l (Ce)*	*Amount of Solute Adsorbed mg(x)*	*Amount Adsorbed Qe = x/m (mg/g)*	$\sqrt{t}$ *(mint)*
60	34.4	15.6	1.56	7.745
180	17.15	32.8	3.285	13.416
360	10.8	39.2	3.92	18.973
720	9.45	40.55	4.055	26.832
1080	8.75	41.25	4.125	32.863

Appendix C

Data for Sugarcane Bagasse

Effect of Adsorbent Dose and Contact Time on Fraction Adsorbed

Contact time	**1/2 hr**
Adsorbent	Raw bagasse
Initial chromium conc.(Ci)	50mg/l
pH	1.5
Size of adsorbent	150 micron
Temperature	30° C

Adsorbent Dose (g)	*Conc. of Cr(VI) after Adsorption, mg/l (Ce)*	*Conc. of Cr(VI) Adsorbed, mg/l (Ci-Ce)*	*Friction Adsorbed [(Ci-Ce)/Ci]*
2.0	41.7	8.3	0.166
4.0	35.5	14.5	0.290
6.0	27.6	22.4	0.448
8.0	22.0	28.0	0.560
10.0	17.6	32.5	0.648
12.0	13.8	36.2	0.724
14.0	10.35	39.5	0.793
16.0	9.0	41.0	0.820
18.0	7.5	42.5	0.850
20.0	7.6	42.4	0.848

Contact time	**1.0 hr**
Adsorbent	Raw bagasse
Initial chromium conc.(Ci)	50mg/1
pH	1.5
Size of adsorbent	150 micron
Temperature	30°

Adsorbent Dose (g)	*Conc. of Cr(VI) after Adsorption, mg/l (Ce)*	*Conc. of Cr(VI) Adsorbed, mg/l (Ci-Ce)*	*Friction Adsorbed [(Ci-Ce)/Ci]*
2.0	38.0	12.0	0.240
4.0	30.0	20.0	0.400
6.0	21.7	28.3	0.566
8.0	16.5	33.5	0.670
10.0	11.4	38.6	0.773
12.0	7.60	42.4	0.848
14.0	5.15	44.85	0.897
16.0	4.0	46.0	0.920
18.0	4.0	46.0	0.920
20.0	4.0	46.0	0.920

Contact time	**2.5 hr**
Adsorbent	Raw bagasse
Initial chromium conc.(Ci)	50mg/1
pH	1.5
Size of adsorbent	150 micron
Temperature	30° C

Adsorbent Dose (g)	*Conc. of Cr(VI) after Adsorption, mg/l (Ce)*	*Conc. of Cr(VI) Adsorbed, mg/l (Ci-Ce)*	*Friction Adsorbed [(Ci-Ce)/Ci]*
2.0	32.5	17.5	0.350
4.0	23.1	26.9	0.538
6.0	15.2	34.8	0.698
8.0	8.5	41.5	0.830

Adsorbent Dose (g)	*Conc. of Cr(VI) after Adsorption, mg/l (Ce)*	*Conc. of Cr(VI) Adsorbed, mg/l (Ci-Ce)*	*Friction Adsorbed [(Ci-Ce)/Ci]*
10.0	4.75	42.25	0.905
12.0	2.20	47.7	0.950
14.0	0.35	49.6	0.99
16.0	0.0	50	1.0
18.0	0.0	50	1.0
20.0	0.0	50	1.0

Influence of pH on Fraction Adsorbed

Adsorbent	**Raw bagasse**
Adsorbent dose	10.0g/l
Contact time	2.5 hr
Initial Cr Conc.(Ci)	50mg/l
Temperature	30° C

pH of the Solution	*Conc. of Cr(VI) after Adsorption, mg/l (Ce)*	*Conc. of Cr(VI) Adsorbed, mg/l (Ci-Ce)*	*Friction Adsorbed [(Ci-Ce)/Ci]*
1.1	0	50	1.0
1.5	4.75	45.25	0.905
2	15.95	34.05	0.68
2.5	31.3	18.70	0.374
3	36.45	13.55	0.271
4.1	44.90	5.10	0.102
5.2	48.0	2.0	0.04
6.3	48.0	2.0	0.04
7	48.0	2.0	0.04
8.5	48.0	2.0	0.04

Effect of Initial Cr (VI) Concentration on Fraction adsorbed

Adsorbent	**Raw bagasse**
Adsorbent dose	10.0 g/l
pH	1.5
Contact time	2.5hr
Size of adsorbent	150 micron
Temperature	30° C

Initial Cr(VI) Conc. before Adsorption, (mg/l)	*Conc. of Cr(VI) after Adsorption, mg/l (Ce)*	*Conc. of Cr(VI) Adsorbed, mg/l (Ci-Ce)*	*Friction Adsorbed [(Ci-Ce)/Ci]*
5.0	0	5.0	1.0
10	0	10	1.0
20	0.84	19.16	.958
30	1.98	28.02	0.934
40	2.88	37.12	0.928
50	4.65	45.35	0.907
60	13.2	46.8	0.780
70	16.1	53.9	0.770
80	20.1	59.9	0.757
90	29.44	60.56	0.757
100	24.3	75.7	0.757

Effect of Contact Time and Particle Size on Fraction Adsorbed

Size of adsorbent	**300 micron**
Adsorbent dose	10.0 g/l
Adsorbent	Raw bagasse
Initial Cr Conc.(Ci)	50mg/l
Temperature	30° C
pH	1.5

Contact Time (hr)	*Conc. of Cr(VI) after Adsorption, mg/l (Ce)*	*Conc. of Cr(VI) Adsorbed, mg/l (Ci-Ce)*	*Friction Adsorbed [(Ci-Ce)/Ci]*
0.5	36.0	14.0	0.28
1.0	27.8	22.2	0.444
1.5	22.45	27.55	0.551
2.0	18.75	31.25	0.625
2.5	15.45	34.55	0.691
3.0	15.0	35.0	0.700
3.5	15.0	35.0	0.70

Size of adsorbent	**150 micron**
Adsorbent dose	10.0 g/l
Adsorbent	Raw bagasse
Initial Cr Conc.(Ci)	50mg/l
Temperature	30° C
pH	1.5

Contact Time (hr)	*Conc. of Cr(VI) after Adsorption, mg/l (Ce)*	*Conc. of Cr(VI) Adsorbed, mg/l (Ci-Ce)*	*Friction Adsorbed [(Ci-Ce)/Ci]*
0.5	31.95	18.05	0.361
1.0	22.4	27.6	0.552
1.5	13.85	36.15	0.752
2.0	8.85	41.15	0.823
2.5	4.45	45.55	0.911
3.0	4.25	45.75	0.915
3.5	4.25	45.75	0.915

Size of adsorbent	**75 micron**
Adsorbent dose	10.0 g/l
Adsorbent	Raw bagasse
Initial Cr Conc.(Ci)	50mg/l
Temperature	30° C
pH	1.5

Contact Time (hr)	*Conc. of Cr(VI) after Adsorption, mg/l (Ce)*	*Conc. of Cr(VI) Adsorbed, mg/l (Ci-Ce)*	*Friction Adsorbed [(Ci-Ce)/Ci]*
0.5	28.75	21.25	0.425
1.0	18.10	31.9	0.638
1.5	9.25	40.75	0.815
2.0	4.825	45.175	0.903
2.5	3.25	46.75	0.935
3.0	3.25	46.75	0.935
3.5	3.25	46.75	0.935

Effect of Contact Time and Temperature on Fraction Adsorbed

Temperature	**30° C**
Adsorbent dose	10.0 g/1
Adsorbent	Raw bagasse
Initial Cr Conc.(Ci)	50mg/1
Size of adsorbent	150 micron
pH	1.5

Contact Time (hr)	*Conc. of Cr(VI) after Adsorption, mg/l (Ce)*	*Amount of Solute Adsorbed mg(x)*	*Amount Adsorbed Qe = x/m (mg/g)*	*Friction Adsorbed [(Ci-Ce)/Ci]*
0.5	31.95	18.05	1.8	0.361
1.0	22.4	27.6	3.02	0.552
1.5	13.85	36.15	3.7	0.752
2.0	8.85	41.15	4.195	0.823
2.5	4.45	45.55	4.50	0.911
3.0	4.25	45.75	4.57	0.915
3.5	4.25	45.75	4.57	0.915

Temperature	**40° C**
Adsorbent dose	10.0 g/l
Adsorbent	Raw bagasse
Initial Cr Conc.(Ci)	50mg/l
Size of adsorbent	150 micron
pH	1.5

Contact Time (hr)	*Conc. of Cr(VI) after Adsorption, mg/l (Ce)*	*Amount of Solute Adsorbed mg(x)*	*Amount Adsorbed Qe = x/m (mg/g)*	*Friction Adsorbed [(Ci-Ce)/Ci]*
0.5	34.6	15.4	1.54	0.308
1.0	24.95	25.05	2.505	0.50
1.5	17.6	32.4	3.24	0.648
2.0	12.52	37.45	3.745	0.749
2.5	9.20	40.8	4.08	0.816
3.0	8.0	42.0	4.20	0.84
3.5	8.0	42.0	4.20	0.84

Temperature	**50° C**
Adsorbent dose	10.0 g/l
Adsorbent	Raw bagasse
Initial Cr Conc.(Ci)	50mg/l
Size of adsorbent	150 micron
pH	1.5

Contact Time (hr)	*Conc. of Cr(VI) after Adsorption, mg/l (Ce)*	*Amount of Solute Adsorbed mg(x)*	*Amount Adsorbed Qe = x/m (mg/g)*	*Friction Adsorbed [(Ci-Ce)/Ci]*
0.5	38.4	11.6	1.16	0.232
1.0	29.40	20.60	2.06	0.412
1.5	22.80	27.20	2.72	0.544
2.0	17.14	32.86	3.286	0.657
2.5	13.54	36.46	3.64	0.729
3.0	13.3	36.7	3.67	0.734
3.5	13.3	36.7	3.67	0.734

Freundlich Adsorption Isotherms

Temperature **30° C**

Weight of the Adsorbent, mg (m)	*Conc. of Cr(VI) after Adsorption, mg/l (Ce)*	*Amount of Solute Adsorbed, mg(x)*	*Amount Adsorbed qe = x/m (mg/mg)*	*Log Ce*	*Log qe*
2000	31.95	17.98	8.99×10^{-3}	1.505	–2.046
4000	22.4	27.29	6.822×10^{-3}	1.356	–2.166
6000	13.85	35.95	5.99×10^{-3}	1.147	–2.222
8000	8.85	41.15	5.143×10^{-3}	0.947	–2.288
10000	4.45	44.97	4.49×10^{-3}	0.701	–2.347

Temperature **40° C**

Weight of the Adsorbent, mg (m)	*Conc. of Cr(VI) after Adsorption, mg/l (Ce)*	*Amount of Solute Adsorbed, mg(x)*	*Amount Adsorbed qe = x/m (mg/mg)*	*Log Ce*	*Log qe*
2000	34.6	15.4	7.70×10^{-3}	1.505	–2.113
4000	24.95	25.05	6.26×10^{-3}	1.356	–2.203
6000	17.6	32.4	5.40×10^{-3}	1.147	–2.267
8000	12.52	37.45	4.68×10^{-3}	0.947	–2.329
10000	9.20	40.8	4.08×10^{-3}	0.701	–2.389

Temperature **50° C**

Weight of the Adsorbent, mg (m)	*Conc. of Cr(VI) after Adsorption, mg/l (Ce)*	*Amount of Solute Adsorbed, mg(x)*	*Amount Adsorbed qe = x/m (mg/mg)*	*Log Ce*	*Log qe*
2000	38.4	11.6	5.80×10^{-3}	1.505	–2.236
4000	29.40	20.60	5.15×10^{-3}	1.356	–2.288
6000	22.80	27.20	4.53×10^{-3}	1.147	–2.343
8000	17.14	32.86	4.10×10^{-3}	0.947	–2.386
10000	13.54	36.46	3.64×10^{-3}	0.701	–2.438

Langmuir Adsorption Isotherms

Temperature **30° C**

Weight of the Adsorbent, mg (m)	*Conc. of Cr(VI) after Adsorption, mg/l (Ce)*	*Amount of Solute Adsorbed, mg(x)*	*Amount Adsorbed qe = x/m (mg/mg)*	$\frac{1}{x/m}$	$\frac{1}{Ce}$
2000	31.95	17.98	8.99 x10^{-3}	111.23	0.0312
4000	22.4	27.29	6.822 x10^{-3}	146.58	0.0440
6000	13.85	35.95	5.99 x10^{-3}	166.90	0.0711
8000	8.85	41.15	5.143 x10^{-3}	194.41	0.1129
10000	4.45	44.97	4.49 x10^{-3}	222.32	0.1988

Temperature **40° C**

Weight of the Adsorbent, mg (m)	*Conc. of Cr(VI) after Adsorption, mg/l (Ce)*	*Amount of Solute Adsorbed, mg(x)*	*Amount Adsorbed qe = x/m (mg/mg)*	$\frac{1}{x/m}$	$\frac{1}{Ce}$
2000	34.6	15.4	7.70 x10^{-3}	129.87	0.0289
4000	24.95	25.05	6.26 x10^{-3}	159.680	0.0400
6000	17.6	32.4	5.40 x10^{-3}	185.185	0.0568
8000	12.52	37.45	4.68 x10^{-3}	213.447	0.0798
10000	9.20	40.8	4.08 x10^{-3}	245.098	0.108

Temperature **50° C**

Weight of the Adsorbent, mg (m)	*Conc. of Cr(VI) after Adsorption, mg/l (Ce)*	*Amount of Solute Adsorbed, mg(x)*	*Amount Adsorbed qe = x/m (mg/mg)*	$\frac{1}{x/m}$	$\frac{1}{Ce}$
2000	38.4	11.6	5.80 x10^{-3}	172.413	0.0260
4000	29.40	20.60	5.15 x10^{-3}	194.174	0.0340
6000	22.80	27.20	4.53 x10^{-3}	220.588	0.0430
8000	17.14	32.86	4.10 x10^{-3}	243.45	0.0583
10000	13.54	36.46	3.64 x10^{-3}	274.27	0.0738

Kinetics of Adsorption

Adsorbent dose	**10.0 g/l**
Adsorbent	Raw bagasse
Initial Cr Conc.(Ci)	50mg/1
Temperature	30° C

Contact Time (hr)	*Conc. of Cr(VI) after Adsorption, mg/l (Ce)*	*Amount of Solute Adsorbed mg(x)*	*Amount Adsorbed Qe = x/m (mg/g)*	*Log (ge - q)*
0.5	32	18.0	1.80	0.442
1.0	19.8	30.2	3.02	0.191
1.5	13.0	37.0	3.70	–0.060
2.0	8.05	41.95	4.19	–0.420
2.5	5.0	45.0	4.50	–1.15
3.0	4.25	45.75	4.57	–
3.5	4.25	45.75	4.57	–

Thermodynamic Parameters

Temperature in Kelvin	*Langmuir Constant*	*Log b*	$\frac{1}{T}$
303	0.2855	–0.5443	3.3 0 x10^{3}
313	0.255	–0.5934	3.19 x10^{3}
323	0.244	–0.612	3.09 x10^{3}

Inter Particle Fiffusion

Adsorbent dose	**10.0 g/l**
Adsorbent	Raw bagasse
Initial Cr Conc.(Ci)	50mg/1
Temperature	30° C
Size of adsorbent	150 micron

Contact Time (mint)	*Conc. of Cr(VI) after Adsorption, mg/l (Ce)*	*Amount of Solute Adsorbed mg(x)*	*Amount Adsorbed Qe = x/m (mg/g)*	$\sqrt{t}$ *(mint)*
30	32	18.0	1.80	5.477
60	19.8	30.2	3.02	7.745
90	13.0	37.0	3.70	9.486
120	8.05	41.95	4.19	10.954
150	5.0	45.0	4.50	12.247
180	4.25	45.75	4.57	13.416
210	4.25	45.75	4.57	14.491

www.ingramcontent.com/pod-product-compliance
Ingram Content Group UK Ltd.
Pitfield, Milton Keynes, MK11 3LW, UK
UKHW021533300726
14060UKWH00011B/487

9 789386 071293